THE WAVE

Also by the Author

The Devil's Teeth

SUSAN CASEY

THE WAVE

IN PURSUIT OF THE
ROGUES, FREAKS, AND GIANTS OF THE OCEAN

Doubleday ~ New York London Toronto Sydney Auckland

DD

DOUBLEDAY

Copyright © 2010 by Susan Casey

All rights reserved. Published in the United States by Doubleday,
a division of Random House, Inc., New York.

www.doubleday.com
www.susancasey.com

DOUBLEDAY and the DD colophon are registered trademarks of Random House, Inc.

Published simultaneously in Canada by Doubleday Canada.

Book design by Maria Carella
Title page photo erikaeder.com
Maps designed by Jeffrey L. Ward
Front endpaper courtesy of enkaeder.com
Back endpaper courtesy of Karsten Petersen, www.global-manner.com

Library of Congress Cataloging-in-Publication Data
Casey, Susan, 1962–
The wave : in pursuit of the rogues, freaks, and giants of the ocean / by Susan Casey.
p. cm.
Includes bibliographical references
1. Rogue waves. I. Title.
GC227.C37 2010
551.46'3—dc22   2010010193

ISBN 978-0-7679-2884-7

PRINTED IN THE UNITED STATES OF AMERICA

3 5 7 9 10 8 6 4

In memory of my father,
RON CASEY

WHEN YOU LOOK INTO THE ABYSS,
THE ABYSS ALSO LOOKS INTO YOU.

Friedrich Nietzsche

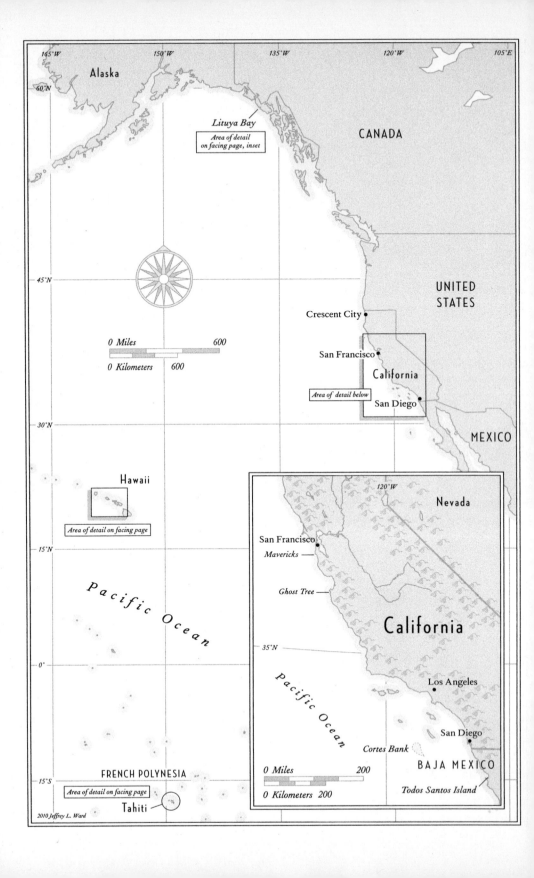

Alaska

165°W   150°W   135°W   120°W   105°E

60°N

Lituya Bay

Area of detail
on facing page, inset

CANADA

45°N

UNITED
STATES

Crescent City ●

0 Miles   600
0 Kilometers   600

San Francisco ●

California

Area of detail below

San Diego

30°N

MEXICO

Hawaii

Area of detail on facing page

120°W

Nevada

San Francisco ●

Mavericks ———

Ghost Tree ———

15°N

California

Pacific Ocean

35°N

0°

Los Angeles ●

Pacific Ocean

San Diego ●

Cortes Bank

BAJA MEXICO

15°S

FRENCH POLYNESIA

Area of detail on facing page

Tahiti

2010 Jeffrey L. Ward

0 Miles   200
0 Kilometers   200

Todos Santos Island

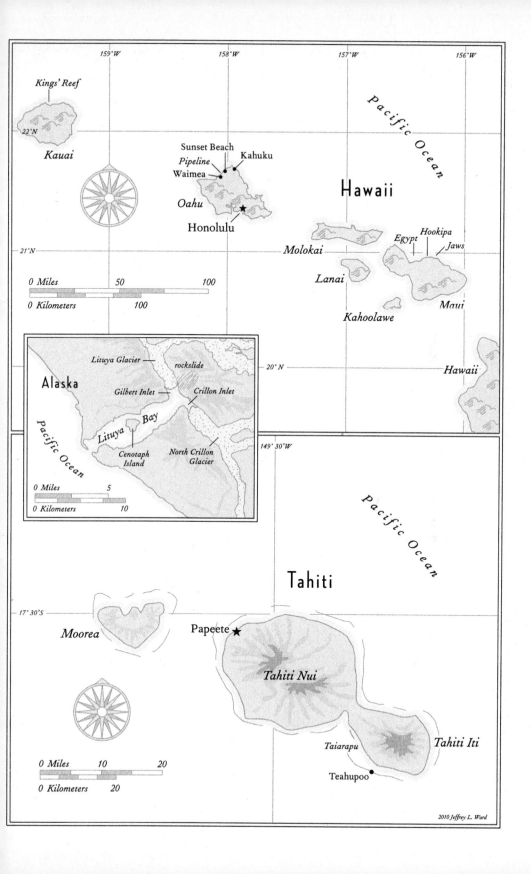

Kings' Reef

22°N

Kauai

Sunset Beach
Pipeline Kahuku
Waimea

Oahu

Honolulu

159°W   158°W   157°W   156°W

Pacific Ocean

Hawaii

Molokai

Egypt Hookipa
Jaws

21°N

Lanai

Maui

0 Miles   50   100
0 Kilometers   100

Kahoolawe

20° N

Hawaii

Alaska

Lituya Glacier   rockslide

Gilbert Inlet   Crillon Inlet

Bay

Lituya

Pacific Ocean

Cenotaph   North Crillon
Island   Glacier

149° 30'W

0 Miles   5
0 Kilometers   10

Pacific Ocean

Tahiti

17° 30'S

Moorea

Papeete ★

Tahiti Nui

Taiarapu

Tahiti Iti

0 Miles   10   20
0 Kilometers   20

Teahupoo

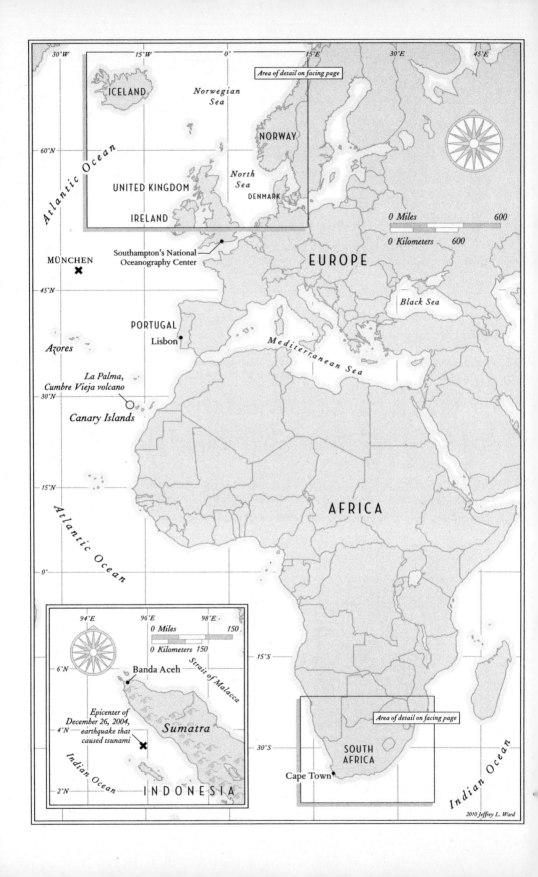

30°W 15°W 0° 15°E 30°E 45°E

Area of detail on facing page

ICELAND

Norwegian Sea

NORWAY

Atlantic Ocean

60°N

North Sea

UNITED KINGDOM

DENMARK

IRELAND

Southampton's National
Oceanography Center

EUROPE

0 *Miles* 600

0 *Kilometers* 600

MÜNCHEN
✖

45°N

Black Sea

PORTUGAL
Lisbon

Mediterranean Sea

Azores

*La Palma,
Cumbre Vieja volcano*

30°N

Canary Islands

AFRICA

15°N

Atlantic Ocean

0°

94°E 96°E 98°E

0 *Miles* 150

0 *Kilometers* 150

6°N
Banda Aceh

Strait of Malacca

15°S

*Epicenter of
December 26, 2004,
earthquake that
caused tsunami*

4°N

Sumatra

Area of detail on facing page

30°S

SOUTH
AFRICA

Indian Ocean

2°N

I N D O N E S I A

Cape Town

Indian Ocean

2010 *Jeffrey L. Ward*

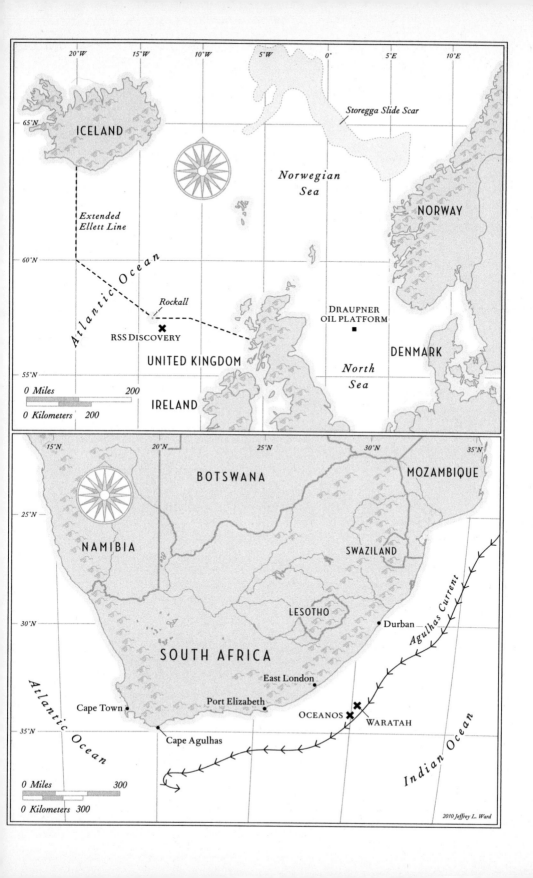

CONTENTS

THE WAVE

INTRODUCTION

The clock read midnight when the hundred-foot wave hit the ship, rising from the North Atlantic out of the darkness. Among the ocean's terrors a wave this size was the most feared and the least understood, more myth than reality—or so people had thought. This giant was certainly real. As the RRS *Discovery* plunged down into the wave's deep trough, it heeled twenty-eight degrees to port, rolled thirty degrees back to starboard, then recovered to face the incoming seas. What chance did they have, the forty-seven scientists and crew aboard this research cruise gone horribly wrong? A series of storms had trapped them in the black void east of Rockall, a volcanic island nicknamed Waveland for the nastiness of its surrounding waters. More than a thousand wrecked ships lay on the seafloor below.

Captain Keith Avery steered his vessel directly into the onslaught, just as he'd been doing for the past five days. While weather like this was common in the cranky North Atlantic, these giant waves were unlike anything he'd encountered in his thirty years of experience. And worse, they kept rearing up from different directions. Flanking all sides of the 295-foot ship, the crew kept a constant watch to make sure they weren't about to be sucker punched by a wave that was sneaking up from behind, or from the sides. No one wanted to be out here right now, but Avery knew their only hope was to remain where they were, with their bow pointed into the waves. Turning around was too risky; if one of these waves caught *Discovery* broadside, there would be long odds on survival. It takes thirty tons per square meter of force to dent a ship. A breaking hundred-foot wave packs one hundred tons of force per square meter and can tear a ship in half. Above all, Avery had to position *Discovery* so that it rode over these crests and wasn't crushed beneath them.

He stood barefoot at the helm, the only way he could maintain traction after a refrigerator toppled over, splashing out a slick of milk, juice, and broken glass (no time to clean it up—the waves just kept coming). Up on the bridge everything was amplified, all the night noises and motions, the slamming and the crashing, the elevator-shaft plunges into the troughs, the frantic wind, the swaying and groaning of the ship; and now, as the waves suddenly grew even bigger and meaner and steeper, Avery heard a loud bang coming from *Discovery*'s foredeck. He squinted in the dark to see that the fifty-man lifeboat had partially ripped from its two-inch-thick steel cleats and was pounding against the hull.

Below deck, computers and furniture had been smashed into pieces. The scientists huddled in their cabins nursing bruises, black eyes, and broken ribs. Attempts at rest were pointless. They heard the noises too; they rode the free falls and the sickening barrel rolls; and they worried about the fact that a six-foot-long window next to their lab had already shattered from the twisting. *Discovery* was almost forty years old, and recently she'd undergone major surgery. The ship had been cut in half, lengthened by thirty-three feet, and then welded back together. Would the joints hold? No one really knew. No one had ever been in conditions like these.

One of the two chief scientists, Penny Holliday, watched as a chair skidded out from under her desk, swung into the air, and crashed onto her bunk. Holliday, fine boned, porcelain-doll pretty, and as tough as any man on board the ship, had sent an e-mail to her boyfriend, Craig Harris, earlier in the day. *"This isn't funny anymore,"* she wrote. *"The ocean just looks completely out of control.*" So much white spray was whipping off the waves that she had the strange impression of being in a blizzard. This was Waveland all right, an otherworldly place of constant motion that took you nowhere but up and down; where there was no sleep, no comfort, no connection to land, and where human eyes and stomachs struggled to adapt, and failed.

Ten days ago *Discovery* had left port in Southampton, England, on what Holliday had hoped would be a typical three-week trip to Iceland and back (punctuated by a little seasickness perhaps, but nothing major). Along the way they'd stop and sample the water for salinity, temperature, oxygen, and other nutrients. From these tests the scientists would draw a picture of what was happening out there, how the ocean's basic characteristics were shifting, and why.

These are not small questions on a planet that is 71 percent covered in salt water. As the Earth's climate changes—as the inner atmosphere becomes warmer, as the winds increase, as the oceans heat up—what does all this mean for us? Trouble, most likely, and Holliday and her colleagues were in the business of finding out how much and what kind. It was deeply frustrating for them to be lashed to their bunks rather than out on the deck lowering their instruments. No one was thinking about Iceland anymore.

The trip was far from a loss, however. During the endless trains of massive waves, *Discovery* itself was collecting data that would lead to a chilling revelation. The ship was ringed with instruments; everything that happened out there was being precisely measured, the sea's fury captured in tight graphs and unassailable numbers. Months later, long after Avery had returned everyone safely to the Southampton docks, when Holliday began to analyze these figures, she would discover that the waves they had experienced were the largest ever scientifically recorded in the open ocean. The significant wave height, an average of the largest 33 percent of the waves, was sixty-one feet, with frequent spikes far beyond that. At the same time, none of the state-of-the-art weather forecasts and wave models—the information upon which all ships, oil rigs, fisheries, and passenger boats rely—had predicted these behemoths. In other words, under this particular set of weather conditions, waves this size should not have existed. And yet they did.

History is full of eyewitness accounts of giant waves, monsters in the hundred-foot range and beyond, but until very recently scientists dismissed them. The problem was this: according to the basic physics of ocean waves, the conditions that would produce a hundred-footer were so far beyond rare as to virtually never happen. Anyone who claimed to have seen one, therefore, was engaging in nautical tall tales or outright lies.

Still, it was hard to discount a report from the polar hero Ernest Shackleton, hardly the type for hysterical exaggeration. On his crossing from Antarctica to South Georgia Island in April 1916, Shackleton noticed odd movements in the night sky. "A moment later, I realized that what I had seen was not a rift in the clouds, but the white crest of an enormous wave," he wrote. "During 26 years experience of the ocean in all its moods I had not encountered a wave so gigantic. It was a mighty upheaval of the ocean, a thing quite apart from the big white-capped seas that had been our tireless enemies for many days." When the wave hit his ship, Shackleton and his crew were "flung forward like a cork," and the boat flooded. Fast bailing and major luck were all that saved them from capsizing. "Earnestly we hoped that never again would we encounter such a wave."

The men on the 850-foot cargo ship *München* would have seconded that, if any of them had survived their rendezvous with a similar wave on December 12, 1978. Considered unsinkable, the *München* was a cutting-edge craft, the flagship of the German Merchant Navy. At 3:25 a.m. fragments of a Morse code Mayday, emanating from 450 miles north of the Azores, signaled that the vessel had suffered grave damage from a wave.

But even after 110 ships and 13 aircraft were deployed—the most comprehensive search in the history of shipping—the ship and its twenty-seven crew were never seen again. A haunting clue was left behind: searchers found one of the *München*'s lifeboats, usually stowed sixty-five feet above the water, floating empty. Its twisted metal fittings indicated that it had been torn away. "Something extraordinary" had destroyed the ship, concluded the official report.

The *München*'s disappearance points to the main problem with proving the existence of a giant wave: if you run into that kind of nightmare, it's likely to be the last one you'll have. The force of waves is hard to overstate. An eighteen-inch wave can topple a wall built to withstand 125-mile-per-hour winds, for instance, and coastal advisories are issued for even five-foot-tall surf, which regularly kills people caught in the wrong places. The number of people who have witnessed a hundred-foot wave at close range and made it back home to describe the experience is a very small one.

Even if a ship does manage to survive a hundred-foot wall of water, there are no underwhelmed survivors. Big fish tales are human nature. Add to that a dose of mortal terror, honest confusion, a fear of being blamed for damage to the ship—if, say, the wave didn't quite measure up to the "something extraordinary" test but managed to poleax the vessel anyway because the captain was below deck playing darts and drinking vodka at the time—and what you've got is less than the scientifically immaculate truth.

But there was a rare occasion in 1933, when a sharp-eyed naval officer aboard the 478-foot oil carrier USS *Ramapo* happened to be up on the bridge as an astonishing wave lurched out of the Pacific and his response, rather than screaming and covering his eyes, was to make a trigonometric calculation using the ship's dimensions relative to the wave's crest and trough. The result was a height estimate that, if not on par with the exactitude of the *Discovery*'s sensors, was at least defensible. And the officer's measurement? The wave was 112 feet high.

If a 112-foot wave isn't freakish enough, consider that this one had leaped out of 45-foot seas. Thus it was more than twice the average size of anything else in the *Ramapo*'s path, which matches the scientific definition for a freak (or rogue) wave. For centuries mariners had spoken of the "hole in the ocean," a cavernous trough at the base of an abnormally large wave, and the "three sisters," a series of freaks in rapid succession. To scientists this kind of folklore was a hard sell. The numbers didn't add up. Maybe, just maybe, a once-in-an-aeon wave triple the size of its surrounding seas might exist—but there was no way traditional oceanography could accept this as a typical occurrence. As for the notion of mutant walls of water showing up in sets, that was not even worth discussing. Then something happened that no one could ignore.

On January 1, 1995, the North Sea was feisty due to a pair of storms, a brutish one crawling northward and a smaller one moving southward to meet it. Statoil's Draupner oil-drilling platform sat somewhere between them, about one hundred miles off the tip of Norway. For the crew who lived on the rig it was a New Year's Day of thirty-eight-foot seas rolling by, as measured by the laser wave recorder on the platform's underside. Unpleasant, perhaps, but not especially dramatic—until three o'clock in the afternoon, when an eighty-five-foot wave came careening over the horizon and walloped the rig at forty-five miles per hour. While the Draupner sustained only moderate damage, the proof was there. This wasn't a case of laser malfunction or too many aquavit toasts the night before. It was the first confirmed measurement of a freak wave, more than twice as tall and steep as its neighbors, a teetering maniac ripping across the North Sea.

They were out there all right. You could call them whatever you wanted—rogues, freaks, giants—but the bottom line was that no one had accounted for them. The engineers who'd built the Draupner rig had calculated that once every ten thousand years the North Sea might throw them a sixty-four-foot curveball in thirty-eight-foot seas. That would be the maximum. Eighty-five-foot waves were not part of the equation, not in this universe anyway. But the rules had changed. Now scientists had a set of numbers that pointed to an unsettling truth: some of these waves make their own rules. Suddenly the emphasis shifted from explaining why giant waves couldn't simply leap out of the ocean to figuring out how it was that they did.

This was a matter of much brow sweat for the oil industry, which would prefer that its multimillion-dollar rigs not be swept away. It had happened before. In 1982 the Ocean Ranger, a 400-foot-long, 337-foot-high oil platform located 170 miles off the coast of Newfoundland, was struck by an outsize wave in heavy weather. We'll never know how big the wave was exactly, for there were no survivors. Approved for "unrestricted ocean operations," built to withstand 110-foot seas and 115-mile-per-hour winds, considered "indestructible" by its engineers, the Ocean Ranger had capsized and sank close to instantly, killing all eighty-four people on board.

In the nautical world things were even more troubling. Across the global seas ships were meeting these waves, from megaton vessels like the *München*—oceangoing freighters and tankers and bulk carriers—down to recreational sailboats. At best, the encounters resulted in damage; at worst, the boat vanished, taking all hands with it. "Two large ships sink every week on average [worldwide], but the cause is never studied to the same detail as an air crash. It simply gets put down to 'bad weather,' " said Dr. Wolfgang Rosenthal, senior scientist for the MaxWave Project, a consortium of European scientists that convened in 2000 to investigate the disappearing ships.

While Rosenthal's numbers may be high, his point is well taken. Given the lack of survivors or evidence, exact statistics of ships scuttled by giant waves are impossible to come by; but it is clear that every year, on average, more than two dozen large ships sink or otherwise go missing, taking their crews along with them. (If you also consider smaller vessels, the numbers are vastly higher.) In particular, a type of ship known as a bulk carrier is vulnerable: on one infamous occasion in March 1973, two bulk carriers were lost within an hour of each other in the same area of the North Atlantic.

When I first read about the missing ships, I was astonished. In the high-tech marine world of radar, EPIRB, GPS, and satellite surveillance, how could hundreds of enormous vessels just get swallowed up by the sea? And furthermore, how could this be happening without much media notice? Imagine the headlines if even a single 747 slipped off the map with all its passengers and was never heard from again.

Clearly, there *was* something extraordinary going on out there. After the Draupner incident, it became undeniable: no one really had a clue as to how waves behaved in their most extreme forms. Yet lives depended on this information. As the scientists scrambled and the oil companies mobilized and the naval architects double-checked their calculations and ship captains worried the horizon, I imagine they thought to themselves: *So the old stories were true after all.*

The first time I saw a truly big wave was in December 1989. I happened to be in Hawaii and my trip coincided with the Triple Crown of Surfing, a series of three competitions held on Oahu's north shore. In order to have the events, though, first you must have the waves. Sometimes the surfers had to wait weeks or even months for the right conditions to materialize, and so it was lucky and unusual that a good-size swell arrived during my visit. On the day the big-wave contest was called at Sunset Beach, I drove my rental car across the island and landed on that stretch of sand, along with about a thousand other people.

The spectator scene was a riot of color, of neon pink bikinis and canary yellow surfboards and lime green banners and all the glimmering blues of the Pacific Ocean. It was a convention of gear-laden trucks, a bazaar of beach hair, from sun-bleached white to drip-dry dreadlocks. The nearest closed-toe shoe was at least twenty miles away. The sky was cloudless but a veil of mist hung in the air from the force of the waves slamming down. At first I found that startling because the Sunset wave itself—the face the surfers would be riding—broke more than a half mile offshore. But then a set rolled in, a pulse of energy that caused several waves to jump up in size. I watched through binoculars as the waves began to build, ominous lumps in the ocean. The water rose and rose until a tiny rider appeared at the top and dropped onto the face as it exploded into a thirty-foot moving cliff. Whenever a wave broke, the beach shook with a little hum of violence.

Standing on shore, I was scared. I'd witnessed avalanches, explosions, tornadoes, wildfires, and monsoons, and I'd never seen anything as intimidating as those waves. For all the gentle images evoked by the name Sunset Beach, in reality this was a different beast. One surf expert described this break as "the entire Pacific Ocean rearing up to unload on your head." On big days at Sunset, people were often swept away by ferocious currents and surges. Watching, I could easily imagine this. What I couldn't imagine was why anyone would willingly insert himself into these elements.

It felt strange to be terrified of the water. After decades of competitive swimming I'm usually more at home in aquatic environments than I am on land. Over the years I'd done assorted damage to myself on solid ground—bruises, bumps, tears, a knee pieced together by titanium screws—but nothing bad had ever happened to me in the water. Then again, I'd never experienced the water in this particular mood. As I watched the surfers launch themselves into the churning ocean and paddle toward the break, I worried for each of them. Their sport seemed more gladiatorial than athletic, like showing up for work each day to grapple with bull elephants.

Which is why, a few years later, I was stunned to see a photograph of a man riding a wave more than twice the size of Sunset, somewhere in the sixty-foot range. The surfer was Laird Hamilton, a six-foot-three, 215-pound twenty-eight-year-old from Hawaii who looked completely at ease inside a barrel as tall as an office building. His blond hair whipped back in the spray; his muscular arms were spread wide for balance as he plummeted down the wave on a tiny board. He had classically handsome features, chiseled and intense, but no fear showed on his face, only rapt focus. Looking at the picture, I didn't understand how any of this was possible.

Since surfing became popular in the mid-twentieth century, faces in the forty-foot range have represented the outer limits of human paddling abilities. Anything bigger is simply moving too fast; trying to catch a sixty-foot wave by windmilling away on your stomach is like trying to catch the subway by crawling. Never mind, though, because even if you could catch it, there would be no way to ride it. Too much water rushes back up the face of a giant wave as it crests, sucking you, the hapless human (not enough momentum), and your board (too much friction) over the falls. So while the most popular surf spots quickly became so overrun that fistfights erupted in the water, all over the world the most impressive waves were going to waste. To Hamilton and his friends, this was unacceptable. The rules had to change, and a new system invented. So they came up with a technique called tow surfing.

Borrowing ideas from windsurfing and snowboarding, they created shorter, heavier surfboards with foot straps, and thinner, stronger fins that sliced through the water like knives. Then they added Jet Skis and water-ski ropes to the mix, using them to tow one another into perfect position at thirty miles per hour. In the optimal spot, just as the wave began to peak, the rider would drop the tow rope and rocket onto the face. The driver, meanwhile, would exit off the back. Using this method, with its increased horsepower and redesigned gear, a surfer could theoretically catch the biggest waves out there. Riding them—and surviving if you fell—was another story.

Hamilton was the test pilot, followed immediately by other surfers and windsurfers in his circle: Darrick Doerner, Brett Lickle, Dave Kalama, Buzzy Kerbox, Rush Randle, Mark Angulo, and Mike Waltze. Nicknamed the Strapped Crew, they experimented on the outer reefs of Oahu and Maui, far beyond the crowds. "No one was there," Hamilton said. "No one had ridden waves this size. It was the unknown. It was like outer space or the deep sea. We didn't know if we were going to come back."

Anything involving giant waves qualifies as a risky pursuit, but tow surfing seemed to invite disaster. The sport's learning curve was a series of hard lessons, and the price of falling was high. It included dislocated shoulders, shattered elbows, and burst eardrums; broken femurs, snapped ankles, and cracked necks; lacerated scalps, punctured lungs, and fractured arches; hold-downs that Brett Lickle described as "sprinting four hundred yards holding your breath while being beaten on by five Mike Tysons." As for stitches, Hamilton "stopped counting at 1,000."

Regardless of its dangers (or maybe because of them), tow surfing's popularity and visibility grew throughout the 1990s, the surfers venturing onto more treacherous waves every year. They tinkered with equipment. They refined their techniques. Working in teams of two—a driver and a rider—they figured out how to rescue each other in behemoth surf. As the stakes got higher and the margin for error got slimmer, a kind of natural selection occurred. Riders who'd glimpsed their own mortality a little too closely drifted to the sidelines. At the other end of that spectrum was Hamilton. Watching him, you got the feeling that no wave was out of reach. The more intimidating the conditions, the more he seemed to thrive in them.

Then in July 2001 a surf impresario named Bill Sharp issued a challenge. "For 2700 years," his press release read, "the Homerian [*sic*] epic known as the *Odyssey* has been associated with beautiful-but-deadly temptresses, forgetful lotus-eaters, and scary, one-eyed monsters. But now thanks to surf wear giant Billabong, it's associated with an even scarier monster: the elusive 100-foot wave." The company, the press release continued, would offer a prize of $500,000 to any man who rode one. This payday was exponentially larger than anything surfing had seen; millions more would come from sponsors in the wake of the triumph. A select group of tow teams would be invited to participate, a crew Sharp referred to as "the Delta Force of surfing."

It was a sexy frontier, defined by a nice round number. Marketing that number was Sharp's intention; he noted that he'd sold the hundred-foot-wave Odyssey contest, originally named Project Sea Monster, to Billabong in less than fifteen minutes. Prone to flourishes of hype, Sharp delivered vivid sound bites: "The Odyssey is Jacques Cousteau meets Evel Knievel meets *Crocodile Hunter* meets *Jackass*," he said. And almost overnight the idea of the hundred-foot wave became the media grail, tow surfing's equivalent of a moon landing.

There were a couple of snags. First, was it physically possible? No one knew how riding a hundred-foot wave might differ from, say, riding a seventy-five-foot wave. As they grow in size, waves increase dramatically in speed and energy. At what point would the forces overwhelm the equipment, or the surfers? "The 100-foot wave would probably kill anyone who fell off it," *Time* magazine wrote. Honolulu's then–ocean safety chief, Captain Edmund Pestana, agreed: "It's a deadly scenario for everyone involved." The trade journal *TransWorld SURF Business* was blunt: "You're asking these surfers to take huge risks for our titillation."

Next, even if a surfer wanted to take his chances, finding the wave was a problem. Although they were no longer considered imaginary, hundred-foot waves were not exactly kicking around within Jet Ski range. Further complicating things, for tow surfing's purposes not just any hundred-foot wave would do. The enormous seas the *Discovery* encountered; the huge freaks that pop up to batter oil rigs—these are unsuitable, despite their great height. Waves that exist in the center of a storm are avalanches of water, waves mashed on top of other waves, all of them rushing forward in a chaotic jumble.

Surfers need giant waves with a more exclusive pedigree. In their ideal scenario, a hundred-foot wave would be born in a blast of storm energy, travel across the ocean for a long distance while being strengthened by winds, then peel off from the storm and settle into a swell, a steamrolling lump of power. That swell would eventually collide with a reef, a shoaling bottom, or some other underwater obstacle, forcing its energy upward and sideways until it exploded into breaking waves. And that's where the ride would begin—far enough from the storm's center to be less roiled and choppy, but not so far that its power was too diminished. This was a pretty tall order. If the ocean was a slot machine, rideable sixty- or seventy-foot waves came along about as often as a solid row of cherries. And the perfect hundred-foot wave? Hit that one and the sirens would go off as everyone in the casino stopped what they were doing to gawk, and the staff rolled in palettes to help you haul away your money.

A surfer who intended to participate in the Odyssey, therefore, would be signing up for a global scavenger hunt. Not only would he have to ride the wave, he'd have to scour the oceans to find it, monitoring the weather's every nuance like a meteorologist, and then show up at precisely the right moment toting Jet Skis, safety equipment, surf gear, and photographers along with him—not to mention a highly skilled partner who didn't mind risking his life when called upon to do so. This was a surfing competition the way the Space Shuttle was a plane. "The Odyssey makes climbing Everest look easy," one British journalist wrote. Regardless, Sharp was undeterred. "I think everybody's ready," he said. "Now, on the giant days, there's no wave that anyone's backing down from."

Millions of years before there was water on the earth, before steam turned into rain turned into oceans, there were giant waves. There were electromagnetic waves and plasma waves and sound waves. There were shock waves from the many explosions and collisions that made our planet's earliest days so lively. Asteroids smacked into it and sent up waves of molten rock, thousands of feet tall. At one time scientists even believed that an enormous wave of this magma, created by intense solar tides, had swung off into space and become the moon.

Although that particular theory is no longer considered true, it points to something that is: waves are the original primordial force. Anywhere there's energy in motion there are waves, from the farthest corners of the universe down to the cells in your eyeball. I wondered if this was why, after eighteen years, I couldn't stop thinking about that day at Sunset Beach. Far from being an abstraction in the ether—like electrical waves, X-ray waves, or radio waves—those thirty-foot ocean waves were a majestic demonstration of the unseen force that powers everything. Catching a glimpse of something that elemental, that beautiful, and that powerful created one inevitable result: the desire to see it again.

The more I read about the mysteries of freak waves, the more jaw-dropping tow-surfing images I saw (and the more inevitable it became that someone *would* ride a hundred-foot wave), the more fascinated I became. New technologies began to reveal startling information. "Ship-Devouring Waves, Once Legendary, Common Sight on Satellite," read the *USA Today* headline on July 23, 2004, describing how radar was now able to measure waves from space: " . . . a new study based on satellite data reveals the rogues are fairly common." "Rogue Giants at Sea: Huge, Freakish, but Real, Waves Draw New Study," the *New York Times* reported in July 2006. "Scientists are now finding that these giants of the sea are far more common and destructive than once imagined, prompting a rush of new studies and research projects."

From a science and technology standpoint, we humans like to think we're quite smart. Over in Switzerland, for instance, physicists are chasing the Higgs boson, a subatomic speck so esoteric that it's referred to as the "God particle." If we're closing in on *this*, how is it possible that only fifteen years ago a force that regularly demolishes 850-foot-long ships was deemed not to exist?

Quite simply: the ocean doesn't subscribe to the orderly explanations that we would like it to. It's a mosh pit of variables, some of which science has considered and others of which it hasn't—because we don't even know what they are. Though we're more informed about the sea now than we were several hundred years ago when mermaids were listed along with sea turtles in Pliny's *Historia Naturalis*, the depths still hold more secrets than anyone can count. And this lack of knowledge affects far more than ships at sea.

Anyone who lives on this planet is utterly dependent on its oceans. Their temperatures and movements control the weather; their destructive—and life-giving—ability dwarfs anything on land. Now that climate change is an accepted fact with unknown consequences, our vulnerability is sinking in. The earth's mean surface temperature (land and oceans combined) is warmer now than at any other time during the past four hundred years, and it continues to rise. In its 2007 report the Intergovernmental Panel on Climate Change (IPCC) concluded that "the ocean has been absorbing more than 80 percent of the heat added to the climate system." As the waters heat up, wind velocity increases; storm tracks become more volatile; polar ice and glaciers melt, causing sea levels to rise. How far will they rise? All we have is a best guess, continually adjusted upward as new (and discouraging) data arrives. In 2007 the sea levels were predicted to rise about 23.5 inches by 2100. In 2009 that number was raised to 39 inches, a level that would displace some 600 million people in coastal areas. (Other scenarios, like the collapse of the Greenland ice sheet, should they occur, would raise the sea level as much as twenty-three feet. For perspective, that would drown most of Florida.) As a result of all of the above and, likely, other factors no one's aware of yet, average wave heights have also been rising steadily, by more than 25 percent between the 1960s and the 1990s. Planetary waves, massive subsurface ocean waves that play a key role in creating the climate, are speeding up as well. The details about what a warmer planet will look like are still coming into focus, but there is one thing our environmental future will clearly hold: a lot of restless water.

If anyone needed a stark preview of the kinds of situations that a stormier, more liquid world might bring, it came on August 29, 2005, when a twenty-eight-foot storm surge from Hurricane Katrina overwhelmed the levees surrounding New Orleans, submerging 80 percent of the city and killing almost two thousand people. (A record-breaking twenty-seven tropical storms formed in the Caribbean that year.) Intense storms are destructive enough on their own, but when the waves hit land the potential for damage goes off the charts: more than 60 percent of the global population lives within thirty miles of a coastline. Then, of course, there are tsunamis, extraordinarily powerful waves caused by underwater earthquakes and landslides. Six years ago the world watched in horror as an estimated hundred-foot tsunami wave erased the Indonesian city of Banda Aceh, home to 250,000 people, in a matter of minutes. Japan, perhaps the most vulnerable nation, has lost entire coastal populations to the waves. In the geological time frame these sudden inundations are hardly isolated events. Over the course of history volatile seas have wiped cities, islands, and even civilizations from the map.

In Waveland, it was as though the scientists aboard the RSS *Discovery* had dropped through a secret trapdoor in a surly but typical North Atlantic storm and into the darkest heart of the ocean: a place where giant waves not only exist but flourish, a place so obscure to us that we're more familiar with the workings of subatomic particles. What *is* out there? What happens in that place? That's what Dr. Penny Holliday and her team wanted to learn. And so did I.

Five years ago I set out to understand giant waves through the eyes of the people who knew them most intimately: the mariners, for whom Shackleton's "massive upheaval of the ocean" is a present and serious threat; the scientists, who are in a race against time to understand the intricate complexities of the sea in a rapidly changing world; and of course, the tow surfers. The members of this rarefied tribe—maybe fifty highly skilled riders across the globe—don't just stumble across giant waves or steer their ships clear of them or consider them as equations on a computer screen, they seek them out. While everyone else goes to great lengths to avoid encountering a hundred-foot wave, these men want nothing more than to find one.

What kind of person drops in on Mother Nature's biggest tantrums for fun? What drives him? And since he has gone into that dark heart of the ocean and felt its beat in a way that sets him apart, what does he know about this place that the rest of us don't? My questions went on, but I knew one thing for sure: if you followed the wave experts into the waves, you would have an interesting—and turbulent—time.

Having wandered some distance among gloomy rocks,
I came to the entrance of a great cavern . . . Two
contrary emotions arose in me, fear and desire—fear of
the threatening dark cavern, desire to see whether
there were any marvelous things in it.

Leonardo da Vinci

THE GRAND EMPRESS

HAIKU, HAWAII

Eight miles east on Maui's Hana Highway, in the shadow of the Haleakala volcano, away from the tourists streaming to the island's lush southern beaches, there is a candy box of a town called Paia. Only a few blocks in size, its streets thrum with locals-only bars, open-air seafood joints, yoga studios, shops selling bikinis and hemp T-shirts and dolphin-themed art. The peace-love-aloha vibe aside, Paia's main purpose is instantly obvious: every vehicle bristles with surfboards.

The surfers are headed to Spreckelsville and Hookipa, nearby stretches of the north shore where the waves are consistently lively. Both areas are wild and exposed; neither is a spot for beginners. Compared to what lies a little farther up the road, however, they're a pair of kiddie pools. The true spectacle requires another five miles of driving, past the blink-or-you'll-miss-it town of Haiku, down a red-dirt path bearing the signs "No Trespassing," "Beware of Dog," and "Authorized Personnel Only," and through a sea of green pineapple fields. At the foot of those fields, there is a cliff.

It's a lonely spot with a harsh beauty, blasted by wind and pummeled by the sea that surges in, three hundred feet below. But a half mile offshore, a number of geological features have combined to create something even more dramatic and foreboding: a giant wave called Pe'ahi, also known by its nickname, Jaws.

For about 360 days a year Jaws lies dormant, indistinguishable from the seas around it, waiting for the right conditions to come along and set it off, like a match to a gas leak. This is one of the first places the North Pacific storms hit, menacing splotches on the radar maps spiraling down from the Aleutian Islands. When a powerful enough storm arrives, all of its energy—which has traveled through water hundreds and even thousands of feet deep—trips on Jaws' fan-shaped reef. Deep channels on either side of the reef, carved by millennia of lava flow and freshwater drainage from the Pe'ahi Valley, above, funnel the energy inward and upward. (Imagine a runaway Mack truck suddenly hitting a ramp.)

The result is sixty-, seventy-, and eighty-foot waves, so beautifully shaped and symmetrical that they might have come from Poseidon's modeling agency. The white feathering as the wave begins to crest, the spectrum of blues from rich lapis to pale turquoise, the roundness of its barrel, the billowing fields of whitewater when it comes crashing down—when you envision the cartoon-perfect giant wave, the gorgeous snarling beast of Japanese landscape paintings, what you are seeing is Jaws.

As far back as the 1960s surfers had been coming to the cliff and eyeballing Jaws. "This is a super freak wave," the famed surfer Gerry Lopez said after one reconnaissance. "Looking at it makes you physically nauseous." Lopez, a 1970s pioneer on some of the Pacific's most fearsome waves, had originally nicknamed Jaws "Atom Blaster," because "it broke like an atomic bomb." That didn't stop people from wanting to ride it, though, and when tow surfing came along, they got their chance. They learned a few things right away. Most important: like all sets of jaws, this one had a tendency to snap shut, swallowing anything unfortunate enough to be inside it. And its teeth . . . well, they were more like fangs.

On a gusty afternoon in late October 2007, I sat in the passenger seat of a battered golf cart as it drove past the Pe'ahi cliff and wound down a steep, stony path toward the ocean. At the wheel was Teddy Casil, a rugged Hawaiian with a bouncer's physique and a don't-mess-with-me vibe. With his left hand, Casil alternated steering the vehicle with drinking a can of Coors Light; in his right hand he held a large machete. Every so often we stopped so he could hack off some jungly tentacle that was blocking our way. At times the path became so precipitous and twisty and thick with red mud that I thought we might just cartwheel to the bottom. But this was no ordinary golf cart. It had been jacked up, fitted with knobby tires, Recaro seats, all-wheel drive, and safety netting. It was ready for anything, its owner made sure of that. And he was right behind us, driving an enormous tractor: Laird Hamilton.

Hamilton, as mentioned, is not the typical small and wiry surfer dude you see on the World Cup Tour, doing flippy tricks in ten-foot waves. He's a large guy, and visibly powerful, a huge advantage in the biggest seas. His back muscles, shaped by decades of paddling, are so defined that they almost seem to push him forward. It is when sitting atop a piece of earth-moving machinery or balanced at the peak of a seventy-foot wave that Hamilton most comfortably fits into scale. Not every successful life seems inevitable, but in this case it's as though fate set out to tailor-make a human being for one specific pursuit. Hamilton's size, his abilities, his mind-set, his upbringing—everything pointed him into the ocean's heaviest conditions.

California-born but Hawaii-bred, he was raised with the planet's most famous surf break—Pipeline—only steps from the house on Oahu's north shore where he lived with his mother, JoAnn, and his stepfather, Bill Hamilton, a star big-wave rider in the 1960s and 1970s. (The story of how three-year-old Laird selected his own father is etched into surf-world lore. His biological father having left the scene shortly after his birth, Laird encountered Bill Hamilton, then a seventeen-year-old fledgling pro surfer, on the beach. The two connected instantly and body-surfed together for an hour or two, the child clinging to the teenager's back. Afterward, Laird told him, "I think you need to come home and meet my mother." Bill Hamilton and JoAnn Zerfas married eleven months later.) And if all that didn't make for a perfect enough petri dish, Gerry Lopez lived next door, acting as a mentor. When Hamilton was six, his father decided to escape Oahu's growing crowds by moving the family to the wilds of Kauai, at the northern tip of the Hawaiian Islands, where the Pacific storms hit first and hardest.

Back then Kauai was a kind of Hawaiian Hades all but closed to outsiders, and Wainiha, the north shore encampment where the Hamiltons lived, was a rugged, isolated backwater where things like electricity and indoor plumbing were scarce. Though it's hard to imagine Laird Hamilton being picked on, his non-native status made school one perpetual fight. Surfing was a way to channel the frustration; by age thirteen Hamilton had become a respected presence at Kauai's most demanding breaks. Between the fierce Na Pali Coast in his front yard and the serpentine rivers that streamed off Mount Wai'ale'ale (a 5,200-foot volcanic peak that has the distinction of being the wettest spot on earth) in his backyard, Hamilton said, "I just happened to grow up in the most aggressive water in the world."

When I decided to head out in search of giant waves, he was the obvious person to call. Our paths had crossed before. During the 1990s I'd worked at a magazine that covered extreme sports, and Hamilton's exploits qualified, to say the least. Over the years I followed his career as it progressed from "Hey, what's he doing?" to "Oh my God, look at what he's doing!" to a level even beyond that, where the most common response was speechless gaping. By the time Hamilton turned thirty he was already hailed as a legend; now, at forty-three, he was still considered the greatest big-wave rider, despite a talented pack of would-be successors trying their best to dethrone him.

Not only did he ride waves that others considered unrideable, at Jaws and elsewhere, but he did it with a trademark intensity, positioning himself deeper in the pit, carving bottom turns that would cause a lesser set of legs to crumple, rocketing up and down the face, and playing chicken with the lip as it hovered overhead, poised to release a hundred thousand tons of angry water. He seemed to know exactly what the ocean was going to do, and to stay a split second ahead of it.

That intimacy, that rare knowledge of what it feels like to be part of an eighty-foot wave—to be in it, to be on it—was something I wanted to understand. So I had come to Maui. This was where tow surfing had been brought to the world's attention, and Jaws was still the gold standard for giant waves. It was also the reason why Hamilton lived on this island, at the top of these pineapple fields: Jaws was literally in his backyard. During a big swell he can feel the wave before he sees it. The ground shakes for miles.

When I'd arrived at his house earlier in the day, Hamilton and Casil were digging a ditch. If the waves were absent Hamilton channeled his energy into working on his land, to tending it and building on it and clearing brush off it. In particular, he loved to move large hunks of it around so that a steeplechase racetrack for golf carts could be created, or a 700,000-gallon pond with a twenty-foot cliff jump carved out of a hillside. Casil, a friend who also helped manage the property, was usually there working with him.

As I stood watching the ditch grow deeper, I noticed a line of steely clouds massing on the skyline. This was typical Maui weather, sudden squalls followed by soft rainbows. In the ocean there were smallish waves coming from the west. But it was almost November, when the Pacific storm swells would begin to arrive, swapping average conditions for threatening ones. Likely Hamilton had that calendar on his mind when he stepped back from his digging and turned to me. His hair, skin, shorts, and boots were all covered in a brownish-red dust. "You wanted to swim out to Pe'ahi?" he said. "Today's a good day."

I did want to do this. After hearing haunting descriptions of the seafloor topography that creates the wave, I was curious to see it. Some people said the reef was shaped like a fan. Others said it was pointed like an arrow and that its apex disappeared into the gloom of the sea. I'd heard talk of a "tongue of lava" down there, which seemed appropriate for Jaws but also fairly sinister. Hamilton's close friend and fellow big-wave rider, Brett Lickle, had described Jaws' seafloor as being riddled with pits and overhangs and caverns. "It's not this beautiful flat thing down there," he said, describing how during a wipeout "there are tons of little holes and places that you can get stuck."

"So it's calm out there right now?" I asked.

Hamilton smirked. "Well, for this time of the year, yeah. About as calm as it's gonna get."

As Hamilton, Casil, and I emerged from the thick vegetation the trail opened up into a cove at the base of the cliffs. Surf heaved in and out against the boulders that ringed its shoreline. The place had an almost northern feel, with fir and pine trees bent at arthritic angles from the wind. There was no hint of the Maui depicted in tourist brochures, nowhere to gradually wade in, no white sand beach. We were two bays up the coast from Jaws, maybe a mile away by water. Casil popped open another Coors Light and set off up the path to do some trail maintenance, followed by Hamilton's two rat terriers, Buster and Speedy, their tails twitching with happiness.

Hamilton, standing in surf shorts and mud-encrusted Wellingtons, gestured toward the water. "Are you ready?" he said. "Do you have your mask? I need you to have good visibility, because we're going to be swimming close to the rocks." As he pulled off his boots, and a rust-colored sock that was once white, I noticed that his right foot was bruised a vivid purple. "The other day I dropped a hundred-pound bench on my foot," he explained. "I broke a toe and dislocated all of the knuckles." He said this in the tone of voice that someone might use to describe a slight irritation, a blister perhaps, or mild sunburn. When you consider what Hamilton's feet have endured—it was. He has snapped his left ankle five times while tow surfing, the joint straining against his foot straps with such force that it finally gave way. One time the bone shattered so sharply that it poked through his skin. He has also broken every toe on his feet (most more than once), fractured both arches multiple times, and lost most of his toenails.

Following him, I edged my way down a tumble of black basalt rocks. Some were slick with red algae that had a ticklish feel. Where ocean met land, the surf swelled and bashed. I watched as Hamilton timed the waves, jumping when one receded but before the next arrived, quickly clearing himself from the impact zone. I looked down. Sea cucumbers and limpets made S-shapes on the rocks. When I saw the whitewater wash back over them, I jumped.

The water was a dusky aquamarine, milky with turbulence. As I adjusted my mask and looked around I saw a field of boulders below, as though we were swimming over a huge upside-down egg carton. It was an elemental place, a seascape of broken rock on an island born from the wrenchings of a volcano. Describing Jaws' surrounding waters earlier, Hamilton said that the wave's intensity made it hard for marine life to thrive anywhere around it. He was right. This was no place for the ornamental or the fragile. The delicate seahorses and cute unicorn fish that floated above reefs on the island's leeward side would last about five minutes in this washing machine.

Hamilton took off in a hail of bubbles. I tried to follow his fins as he threaded through the rocks, but waves tossed me around and I lost sight of him immediately. I steered away from the shoreline to get my bearings. Hamilton's snorkel popped up for an instant and then vanished again beneath a whitecap. For him, swimming out to Jaws on a day when it wasn't breaking was like taking a boat tour of Niagara Falls after you'd already gone over it in a barrel, a deep anticlimax. For me, on the other hand, it was a combination of fear and fascination, the feeling you'd get if you peered into a monster's den while it was asleep.

We headed diagonally across the bay. After a few hundred yards Hamilton stopped and pointed down: "See that hole? That's a miniature version of what's on the reef." Below us lay a maze of rocks; some rounded, some flat, some with sharp, angular corners. They were heaped together in a brutal mosaic, with thin paths snaking between them. In the center was a darker crevice, about the width of a human body.

Jaws' epicenter lay a half mile ahead, but already I could sense that we were in the neighborhood. The water turned abruptly from marine blue to navy-black as the bottom dropped off. Against the darkness it was easy to envision the hazy outline of a tiger shark, its stripe pattern almost a shadow on its massive body. I would have preferred to sprint across this section, but Hamilton stopped and raised his mask. He gestured to some cruel-looking rocks offshore. "A lot of guys wash up on these rocks. See, there's a piece of rescue sled." I looked and saw a white shard jutting up like a dagger, a remnant of the six-foot-long sled that connects to the back of the Jet Ski. Over the years dozens of surfboards, rescue sleds, and Jet Skis had met their end on those rocks, as acres of whitewater boiled toward the cliff. Every forward escape route dead-ended here; anyone stuck nearby would be powerless to avoid the collision. I had always known this was a serious place. But at that moment, seeing the wreckage, it hit me in a visceral way. There were just so many things that could go wrong out here.

It is impossible to think about Hamilton—and Jaws—without figuring Dave Kalama, Darrick Doerner, and Brett Lickle into the picture. Emerging from the larger Strapped group as a tighter unit, the four men shared two key traits: extreme competence in mammoth surf, and a willingness to perform rescues, no matter how dicey the situation. These things were critical because, above all, tow surfing was a team sport. Any surfer who fell at Jaws wasn't getting out of there alone. There was a brief window, maybe a fifteen-second interval between waves, in which a driver had to sight his partner's head in the churning foam, dart in on the Jet Ski, grab him, and get out before the next wave came hurtling down. (Along with its size, Jaws moves with uncommon velocity, approaching forty miles per hour.) It soon became clear that not everyone was up to the task. People froze on the sidelines, or pretended to be very busy elsewhere while their partners floundered in the impact zone. "There were the guys who would come get you and the guys who wouldn't come get you," Hamilton said. "And there was a separation, a big gap, between the getters and the non-getters." Hamilton, Kalama, Doerner, and Lickle were concerted "getters," rescuing anyone who needed help, even surfers they didn't know or whose boneheaded actions had virtually guaranteed a fall.

Kalama and Lickle had begun their wave-riding careers as windsurfers at Hookipa, an exposed stretch of ocean just a few miles from Jaws. For Kalama it was a homecoming; his father's family is one of Hawaii's oldest and most respected, and though he'd been raised in southern California and had a successful stint as a ski racer, Maui called him back. With his curly blond hair and green eyes, Kalama didn't look much like a native Hawaiian, but right from the start he surfed like one. In short order he mastered windsurfing, then expanded his repertoire to include surf canoeing, outrigger paddling, standup surfing, and of course, tow surfing. Kalama was softer spoken and slightly less physically imposing than Hamilton, but of all the men he came closest to equaling him in the waves.

Lickle was from Delaware. At twenty-one he'd come to Maui on a vacation, decided he'd found his ideal place, and vowed to return for good. Back east he registered his intentions by rigging a windsurfer in his bedroom and hanging from the harness for hours at a time. Tough, funny, and burly, he'd established himself in 1987 by windsurfing a fifty-foot wave on Maui's north shore that was, at the time, the biggest anyone had ever ridden.

For all the craziness of his chosen profession, however, the forty-seven-year-old Lickle had learned the meaning of caution. Over time people's horrendous injuries and his own near misses had taught him that even the best get unlucky. He believed in instinct, intuition, and the wisdom of listening to that faint, whispering voice in your head when it advises you to stay onshore. "Sometimes if it doesn't feel right, I'll put my board right back in the car," he said.

Doerner lived on Oahu, one of the most venerated lifeguards on that island's north shore. Double D, as he was known, had plucked hundreds of people out of seething ocean conditions. While others locked up in panic, Doerner reacted in the opposite way, becoming calmer and more intensely focused under duress. This ability earned him a second nickname: the Ice Man. Even in the pre-towing days, he was an accomplished big-wave surfer. He and Hamilton met on Oahu in the 1980s, bonded over their desire to ride even more formidable waves, and then conducted their first towing experiments on the outer reefs beyond Sunset and Pipeline.

The four men adhered to the Polynesian concept of the "waterman," a code that required a surfer to be as all-around confident in the ocean as he was on land. The modern prototype was Duke Kahanamoku, the Hawaiian Olympic swimming champion who also introduced surfing to the world in the 1920s. Like Duke and the Hawaiian kings before him, a true waterman could swim for hours in the most treacherous conditions, save people's lives at will, paddle for a hundred miles if necessary, and commune with all ocean creatures, including large sharks. He understood his environment. He could sense the wind's subtlest shifts and know how that would affect the water. He could navigate by the stars. Not only could he ride the waves, he knew how the waves worked. Most important, a waterman always demonstrated the proper respect for his element. He recognized that the ocean operated on a scale that made even the greatest human initiative seem puny.

Not to behave with humility at Jaws, therefore, was the ultimate karmic sin. "As soon as you think, *I've got this place wired. I'm the man!*," Lickle said, "you're about thirty minutes away from being pinned on the bottom for the beating of your life."

All of them, even Hamilton, had survived rag-doll wipeouts on massive faces. They knew what it felt like to be pummeled by the wave, come to the surface, and then be efficiently whisked to safety by a partner who had his act together. That feeling was far more poignant than mere relief. "You come away and you've cheated something," Lickle said. "I don't like to say death but it's true. It's like you've been given another ticket." To Hamilton's mind, the real peril in falling wasn't physical, even in the case of fatal injury: "You wouldn't even know. It'd be the people you left behind." His deepest fear, he said, was not death but rather "being pounded so bad that psychologically you don't recover."

It was New Year's Day 2000 when this almost happened to Dave Kalama. Jaws was pumping out fifty-foot waves, and Kalama was feeling aggressive. "I was thinking, 'I am just gonna tear this place up today,'" he said. His usual partner, Hamilton, was off the island, so Lickle had towed him into three gorgeous, glassy waves. Then: the fourth. This wave was an ugly stepsister, its face studded with bumps. When Kalama hit one the wrong way he found himself flying backward, looking up at the curling, menacing lip. He remembers thinking, *This is going to get interesting.*

Sucked over the falls, the most disastrous place to be, he caught a flash of blue sky before being slammed down and driven thirty feet deep. Panicking burns oxygen, so he tried to stay calm, tucking in his arms and legs as the wave released its energy, then making for the surface. He was inches from getting a breath when the next wave hit, pinballing him back into the depths. Two wave hold-downs were serious. *This might be it,* Kalama thought, *but let's see.*

When the second wave released him, he broke the surface and saw Lickle nearby. Kalama grabbed the rescue sled, but another mountain of water was already upon them. When it hit, the Ski was sucked backward into a whitewater hole, and Kalama was ripped off the sled and thrust down again, even deeper this time: "I could feel it by the pressure in my ears."

Whitewater blocks out the light, so below the surface everything was black. Kalama, exhausted and disoriented, didn't know which way was up. He began to convulse, his body straining to take a lungful of water while his mind was still barely able to prevent it. Later, he would be told that this is the first stage of drowning.

By luck or skill or grace he resurfaced, and again Lickle was there. Kalama made a desperate lunge for the sled. But Jaws wasn't done with him yet—another wave exploded on top of them and sent the Ski tumbling. "We were rolling underwater," Kalama said. Lickle's feet smacked Kalama's head, but both men held tight and in thirty seconds they were back in calmer waters. "Kind of a rough way to start the new century," Kalama said. "It was baby steps to build my confidence back up. It took me three years to feel like I was in control again."

"That's Jaws beach," Hamilton said, treading water and pointing toward the shore.

I could make out a small, crescent-shaped indentation about eight hundred yards away, filled with rocks. More than that, I could hear it. As the waves swept in and out, the rocks rolled forward and backward, making a sound like an avalanche of bocce balls. It was a rasping, raking noise that was frankly terrifying. I'd read that the ancient Hawaiians considered this a sacred place and held ceremonies on the cliffs above. I could see why. They believed that every last stone and leaf and flower and drop of water contained a spiritual life force, called *mana*, as surely as people and animals did. All things in nature were fully alive. If you shut your eyes and listened to the rocks clacking and grinding, it was as though Pe'ahi had a voice.

We swam on. As we approached the mouth of Jaws, the bottom features changed from midsize rocks to slabs and shelves and monoliths, an aquatic Stonehenge. Here, then, were the molars (and some pointy incisors). The reef was larger than I'd expected—to make out its shape you'd need an aerial view—and also starker, meaner, and more forbidding. Beneath its blue surface, Jaws was a study in grays: slate gray, gray-black, teal-gray, a pale whitish gray. Part of its eeriness, I realized, came from the ghost town atmosphere: there wasn't a fish to be seen. Usually when you're swimming around rocks, you can look down and pick out creatures everywhere. Not here.

I looked around for Hamilton and couldn't find him. There was an instant of panic, and then something flashed below me. Hamilton had dived to the seafloor—forty feet down. I could see his blond hair, brilliant against the gloom. Floating in the swells, I watched as he wound through tunnels and between rocks for what seemed like an aeon. Once I had asked him how long he could hold his breath underwater, figuring this was something he practiced. "There's a school of thought that says you don't train for what you don't want to happen," he replied. "I don't want to consciously know how long I can hold my breath. I just know that so far—long enough."

Hamilton resurfaced, holding a handful of the bottom. It wasn't a fine-grained sand but rather a rough mixture of broken stones. Jaws is not the kind of place that invites lingering, and we turned to head back. The afternoon had ebbed and the water took on an even blacker cast as the sun slipped behind the cliffs. Above us, the gnarled silhouettes of wind-bent trees stood out in sharper relief. The waves were choppier now, the wind angrier. Hamilton stroked toward the rocks, the tightest line available.

I decided to take a longer route to stay away from the rocks, as that made it less likely I would be dashed against them. As I swam, I tried to calm my nerves. There is nothing more unsettling than being alone in a spooky patch of ocean. When three large gray fish darted in front of me, I reared up like I'd been attacked.

Back near the spot where we'd jumped in, Hamilton waited fifty yards offshore. Waves were now exploding against the rocks. "So you'll follow me in," he said. It wasn't a question. We both knew the only way I'd make it onto shore in one piece was to suspend my judgment and do exactly as he said. When he said go, I needed to go. If I hesitated because I wasn't sure his timing was right, I would pay. Judging waves, knowing the pulse of their energy—this was as obvious to Hamilton as any of his five senses. Looking over his shoulder at the incoming surf, he waited until a set had passed, and then he shouted, "Now!" and hightailed it in, exiting the water in a single fluid motion. I hung back an instant too long, got rolled by whitewater, scraped the rocks, and bloodied both my knees.

The house where Hamilton lives with his wife and daughter presides over the pineapple fields with a low-slung, minimalist grace. It is a two-story house, planned along horizontal lines. The living area is upstairs, while the lower floor is given over to a gym and a sprawling garage that, like an airplane hangar, opens on both ends. For Hamilton the garage serves as a combination clubhouse, mission control, and storage facility. Under its roof there are many vehicles, including two old army dump trucks, a trio of souped-up golf carts, three heavy-duty Ford pickup trucks, a Range Rover, a half-dozen Honda Jet Skis on trailers, and a Yamaha jet boat. There are also mountain bikes, road bikes, kids' bikes, a tandem bike, off-road skateboards, a picnic table, two refrigerators, a restaurant-grade espresso machine, and every tool imaginable; shelves filled with generators, shop vacs, gas cans, chain saws, hacksaws, and band saws; and of course, racks and racks of surfboards. Hamilton estimates that he owns about 140 boards, ranging from sleek six-foot tow boards for riding Jaws to majestic twenty-six-foot standup boards for doing things like paddling through the entire Hawaiian Island chain.

To witness the garage—Daredevil Central—is to wonder what Hamilton's wife makes of it all. But anyone who has met Gabby Reece instantly gets the answer. At six foot three, with blond hair down to her waist and an athletic résumé that includes playing NCAA volleyball and professional beach volleyball, and being the first woman to have an eponymous Nike shoe, Reece stands eye to eye with Hamilton on all matters. The two first met in 1995 when Reece, host of a television show called *The Extremists*, invited him on as a guest. The pair skydived together. They married two years later, in a canoe on Kauai's Hanalei River. In 2003 Reece gave birth to their first child, a girl named Reece Viola Hamilton, and that fall she was seven months pregnant with their second. (Hamilton also had a daughter, Izabela, thirteen, from a previous marriage.) Given her own sports background, Reece not only tolerated Hamilton's unusual lifestyle, she supported it wholeheartedly. "It's who he is," she'd said. "You couldn't live with him if he wasn't doing it."

Poised next to Jaws, raising a family, compromising nothing: it had taken more than two decades of hard striving for Hamilton to get to this place. From the start he had turned his back on professional surfing competitions, with their judging panels and sponsorship obligations, and focused his attention entirely on giant waves. This was a noble stance, perhaps, but a decidedly noncommercial one, at least in the beginning. A sole sponsor, the French sports company Oxbow, had supported him since his early days, and Hamilton's loyalty to them ran deep. He and Kalama had also partnered in a film production company, releasing movies each year of their big days at Jaws. As Hamilton's visibility and notoriety grew—and as tow surfing captivated the mainstream—companies like American Express and Toyota came calling. Building a lucrative career had required him to hack his way down a singular path: one, in fact, that hadn't existed before him.

Evening had closed in by the time we returned from our swim, pulling in front of the house next to Hamilton's two three-hundred-pound razorback pigs, Ginger and Marianne. The pigs were snuffling around, gouging divots of mud and grass. Hamilton parked the tractor, walked around the side of the house, and picked up a hose to rinse himself off. Casil disappeared into the garage. I stood looking at the fields as they turned from green to gold, and at the ocean beyond. It's one thing to be told that something is magic, I thought, and another to sense that yourself. It is the difference between seeing a picture of a thunderstorm and finding yourself in the middle of one, smelling the water in the air as the light drains from the sky, hearing the thunder. I definitely wanted to see Jaws when it broke, but even now I was beginning to understand what made the wave unique.

Hamilton walked the hose across the grass and began to wash the mud off Ginger and Marianne. "Do different waves have different personalities?" I asked.

"Absolutely," he said quickly, then hesitated. "Pe'ahi is . . . hmmm . . . the Grand Empress." Generally Hamilton was an articulate person, and when he was passionate about something, he spoke in a rush of words. His voice had a gravelly baritone edge, not a growl exactly but getting there. When he talked about Jaws, though, his thoughts were carefully measured, his tone softer. "Just the magnitude, the sheer volume, the size of the wave, the shape of it," he said. "And it's finicky too. On any given day she'll give someone a kiss and somebody else a slap. You hope you're the one getting the kiss. But she's sensitive that way." He paused for a beat, and then laughed. "I've gotten a spank or two, but not that often. I'm real polite to her."

Not everyone could say the same thing.

As tow surfing headed into its second decade; as it became clear that a person could drastically change his fortunes by having his photograph taken on a seventy-foot wave (with the image zipping around the globe that same day); as wave-forecasting services sprang up so that epic conditions were no longer a local secret—a new cadre of riders was showing up on the biggest days. They were more aggressive than experienced, more brash than respectful. They hadn't spent years honing their skills and practicing rescues and cultivating their partnerships. Because of that, they were dangerous.

The problem came to a head on December 15, 2004. It should have been one of the best days ever at Jaws, but instead the problems began early. "When we got there the first thing I saw was a body skipping down the face," Dave Kalama recalled. In past years there might have been ten tow teams out, all of them familiar players, treating the situation with gravitas. On that morning the scene that greeted the men looked like something out of Fellini's aquatic circus.

Two thousand people lined the cliff, while below the water teemed with photographers, surfers, Jet Skis, and boatfuls of gawkers. At least forty tow teams were buzzing around, and a swarm of other vessels bobbed in the channel next to the wave. Helicopters circled overhead. Many of the top big-wave riders in the world had come to Maui for this swell, but so had dozens of surfers whose best credentials were that they could get their hands on a Jet Ski and find someone to drive it.

People had worried that the hundred-foot-wave prize (the Odyssey had morphed into an event called the Billabong XXL) would lead inexperienced riders into situations that were over their heads, and the craziness of that day seemed to prove them right. Medevac helicopters hoisted out a steady stream of the injured. Jet Skis lay smashed on the rocks. One surfer took such a beating in the whitewater that his flotation vest, rash guard, and trunks were torn from his body, and he lay naked and bloody on the rescue sled as he was driven back to the channel.

Kalama was stunned. "They're going straight to the Indy 500 as soon as they get their drivers' licenses," he said. Lickle was amazed: "I watched guys take off on a sixty-footer, no skill whatsoever. Whole thing hammers them on the head. They take another five waves on the head and then get back on the Ski and do the same thing over again. *What's that about?*" Hamilton was furious. When a Jet Ski had crossed directly in front of him as he dropped into a wave, he was forced to straighten out and surf directly into the impact zone. The violence of the crash split his lips open.

Everyone agreed that half the field wouldn't have come if not for the potential prize money. When the Odyssey had first been announced, Hamilton, Kalama, Doerner, and Lickle made it clear that, far from scrambling to win the thing, they wanted nothing to do with it. "It's all about people wanting to box it up," Hamilton said, angrily. " 'So-and-so rode the hundred-foot wave.' That's by chance. I don't want by chance. I want more performance. What are you *doing* on this hundred-foot wave that you're supposedly riding? Are you running for your life on the shoulder? Are you barely making it? Or are you ripping it apart like it's a twenty-foot wave?" Besides, he added, it was stupid to judge a wave's intensity by height alone. A thick, pugnacious shorter wave could be far more extreme than a tall, anemic one: "Would you rather be attacked by a pit bull or a Great Dane?"

Here was the weird thing: after a decade of churning out at least two humongous days each season, since December 15, 2004, Jaws hadn't broken at anywhere near peak size again. Two winters had passed, a third was beginning, and still Jaws hadn't roared. It was as though the Grand Empress had decided to punish the entire court for misbehavior.

Nothing was more depressing for a big-wave rider than to have months go by when the waves went elsewhere. He felt a sense of purposelessness, frustration, and even depression, the kinds of things you'd feel if you were a mountaineer stuck on the plains, or a Formula One racer in a world that had only Ford Escorts. Hamilton's response was to train even harder, to physically exhaust himself by working outdoors or riding his mountain bike up the volcano or going on long, arduous paddles down the coast. "The busier I stay, the better," he said. "I'm here in the firehouse, waiting for the fire bell to ring."

Eight days later it did.

WAVES ARE NOT MEASURED IN FEET AND INCHES,
BUT IN INCREMENTS OF FEAR.

Big-wave surfer Buzzy Trent

BROKEN SKULLS

PAPEETE, TAHITI

At one o'clock in the morning on October 30, 2007, the Faa'a airport in Papeete, capital of the island nation of Tahiti, was packed. Musicians in Polynesian shirts serenaded the newly arrived visitors, while smiling women in long red dresses handed out white tiaré flowers. The flowers were tiny but their fragrance filled the air, which felt hot and sticky even in the middle of the night. At the baggage claim, things were chaotic. Hundreds of padded surfboard bags and hard-sided camera cases and oversize duffel bags and suitcases and crates and boxes were brought out until the area couldn't hold any more, and trucks pulled up in front of the little open-air terminal to load the cargo, jamming the road. Horns honked, people yelled. Everyone jockeyed with too much gear, and a frazzled energy ran through the place like a current.

The crowd was composed almost entirely of men, most of whom knew one another. Though they had flown to Tahiti from all corners of the globe, they could always recognize the members of their tribe. For one thing, they had a uniform: long, low-slung shorts, flip-flops, hoodies, and T-shirts emblazoned with logos for Quiksilver and Billabong and Hurley and Pipeline Posse. Most wore a hat of some sort, a baseball cap or a woolen ski cap pulled low over the ears. Among these brethren there were no potbellies or thick glasses, no pasty complexions cultivated under fluorescent lights. It was a sea of tans, tattoos, testosterone, and nerves stretched tight as wire.

The men had all come here for the same reason: a legendary wave called Teahupoo (pronounced tay-ah-HOO-po). About twenty-eight hours from now a gigantic swell was due to arrive at these shores, causing the wave—its Tahitian name translates loosely into "Broken Skulls"—to erupt into its full, feral splendor. No one was more identified with old Broken Skulls than Laird Hamilton, who strode through the airport unconcerned that his presence rattled every other surfer in the terminal. If Hamilton was here, they all knew, the waves would be serious.

Like Jaws and other spots where the right storm brought not only big waves but also a life-and-death proposition, Teahupoo was known for its special brand of jeopardy. It was a mean vortex with a deep belly and a thick slab of a lip that all but promised to pile-drive the surfer into the reef, which lay only a few feet below the surface. So much water and energy exploded in such a compressed area that oceanographers often referred to Teahupoo's hydraulics as "freakish." It terrified even the most seasoned big-wave riders—and that was on a small day.

Then there were the not-so-small days. In particular there was August 17, 2000. The only three words on the *Surfer* magazine cover that featured a photo of Hamilton riding Teahupoo on that date were "Oh My God." Only a few months earlier a surfer had died in far tamer conditions, his neck broken and face torn off on the jagged coral, and that must have been on everyone's mind when Hamilton, towed by Darrick Doerner, took off on a wave so massive and so vicious that spectators watching from boats feared they were watching a man's last ride. There was just so much fury in this wave, a meat-grinder-cement-truck-wrath-of-God fury, that even now to watch it on video is a sobering experience. As the wave rose around him Hamilton couldn't see it; he was riding backside, his body turned away from the barrel. But he could feel it, and his mind, he said later, screamed at him to abort, eject—anything but go through with this. At the same time, hesitation would have been fatal. As the lip slammed down on the reef the tube convulsed, the spray exploded around him, and he disappeared from sight. For a heartbeat or three no one knew if the wave had gotten him. Then he emerged, gliding along with his arms in the air. If Hamilton had fallen, the wave cognoscenti agreed, the only thing left of him would have been a red stain on the reef.

The details of that day, and the degree to which Teahupoo had demanded that Hamilton walk the razor's edge of his abilities in order to survive it, had been written into history as "the heaviest wave ever ridden." As if to emphasize the memory, the approaching storm—an out-of-season southern hemisphere low that had brewed in Antarctica before winding toward the South Pacific, fueling itself with tropical moisture along the way—was shaping up to be the hardest hit in years. It was fierce enough, slow-moving enough, and visible enough on the weather maps to give the surfers a full two days' notice to get to Tahiti. Hamilton had been eating a breakfast of ahi and eggs in Maui when he'd gotten the call from his friend Raimana Van Bastolaer, a Tahitian surfer, the previous morning. Thirty-four hours later he landed in Papeete.

I found my bags and lugged them to the curb, where I waited for Hamilton and the two photographers who accompanied him, Sonny Miller and Jeff Hornbaker. Miller, forty-seven, and Hornbaker, forty-eight, were well-known names in the world of surf imagery. Both men were Californians who had developed a taste for the waves early in life. Miller, wild-haired, blue-eyed, and given to an infectious, sandpapery laugh, was a former pro skateboarder known for ocean cinematography that was half fine art, half gonzo mission. Hornbaker was a tall, soulful guy with white whiskers and sun-blasted skin; his work was iconic, beautiful, and principled. In the beginning Hornbaker shot nothing but empty waves, mesmerized by their form. As a purist nothing vexed him more than crass sellouts, or people who failed to see the simple magnificence in the world around them.

In the crowd I caught sight of Sean Collins, the founder and chief surf forecaster of Surfline.com. Above anyone, he was aware of how the swell bearing down on Tahiti was progressing; if pressed, he probably could have recited its real-time longitude and latitude. Collins, fifty-four, was a wave-forecasting savant, and he had turned that talent into a formidable business. Surfline, with its slogan "Know Before You Go," provided guidance to anyone with a vested interest in what the waves were up to. The service delivered forecasts, weather maps, wave models, news, stories, photo galleries, videos, webcams, travel information, and glossaries; it did everything but wax the surfboard for you.

Yet as all big-wave surfers knew painfully well, no forecast was infallible. For every storm that appeared right on cue, at the size and intensity that was predicted, there were two that defied expectations. Small swells turned out to be nasty (and not so small) backhand slaps; lumbering beasts came in with a whimper. Occasionally swells popped out of nowhere, their very existence unforeseen. The best long-range bets were storms that, like this one, showed up as such a pronounced whorl on the radar maps that even the worst-case scenario would produce noteworthy surf. Which was a comforting thought if you had just traveled halfway around the world in search of that.

Collins, a Pasadena-born surfer himself, had only a hint of the detail-obsessed weather geek to his appearance. It was in his eyes, which scanned the airport taking everything in, as if logging data for a future computer program. Remove that, and you were left with a pale-haired guy of average height, with a shy smile and a low, quiet voice, standing around in jeans and a T-shirt. I was curious to hear what he had to say about this storm, so I went over to ask. "It's going to be big," he said, with a nod on the final word for emphasis. "It's a dangerous swell, for sure. The storm is close. It's within two and a half days away. The satellite at that location measured fifty-foot seas and sixty knots of wind." Collins paused, looking out at the horde of surfers. "Everybody's here," he said, then quickly revised that. "Well, some guys didn't want to come. One said, 'I've got a kid on the way. I don't need to split my head on the reef right now.'"

At that moment Hamilton walked over to steer me toward the rental car. Seeing that I was talking to Collins, his face turned stony. "Well, Sean," he said, "I see you've got the whole world here." Being on the receiving end of Hamilton's anger is not pleasant for anyone, and Collins flinched beneath his baseball cap. Without waiting for a response, Hamilton picked up my bags, turned, and walked away. I followed. "This is what happens when you send out a mass e-mail that says, 'Giant Swell!'" he said, waving his hand toward the terminal.

Unlike many of the surfers here, Hamilton was old enough to remember the days when you didn't have forty-eight hours' advance notice to get to the waves, and had to rely on a combination of patience, persistence, luck, and your own meteorological divinations to score an exceptional ride. No one e-mailed you the information; that expertise couldn't be bought. Now it was on sale for a monthly subscription fee, along with all the other gear a fledgling tow surfer might want. Like it or not, the days of having to sniff the air for signs of a storm and invent the equipment in your garage had been overtaken by technology and a thriving commercial enterprise.

Miller and Hornbaker were busy loading a dozen cases of camera gear into a pickup truck that was already piled high with Hamilton's surfboards. The truck would meet us at Raimana Van Bastolaer's house, where we were staying. For a little island, Tahiti had more than its share of renowned big-wave riders, and Van Bastolaer was one of the best. Small-statured and agile, he began his career at Teahupoo as a body boarder, braving the waves with only a pair of fins. (Instead of a surfboard slicing through the water, a body boarder uses, well, his body.) This activity was confined to the tamer days, of course, but it allowed Van Bastolaer to learn the nuances of the wave's motion; by the time he was tow surfing it on the biggest days, he knew its every last trick. He had home court advantage and he surfed it that way, but he also made sure the visiting teams had everything they needed. When a swell rolled in and the megawave *Who's Who* came to town, Van Bastolaer could be found in the center of it, an omnipresent host. He was acquainted with all the players—the locals and the outsiders, the sponsors and the riders—and he coordinated things to such an extent that surfers often referred to Tahiti as "Raimana World." On this swell, he and Hamilton would be tow partners.

The pickup fully loaded, Hamilton, Miller, Hornbaker, and I squeezed into our rental car, a black boxy subcompact that was apparently the last available vehicle on the island. "Like the car?" Hamilton said, wedging himself into the driver's seat. "I could probably lift it." He steered through the crowds, still overflowing from the baggage area onto the road. Rolling down the window, Miller let loose with a long, rattling laugh. "Everybody's coming to Tahiti," he said, leaning outside. "Look at this madness!"

Three hours later I awoke to clanking noises and mumbled voices. Groggy, lying on a mattress in the dark, hearing waves breaking nearby, I experienced a brief, disorienting moment of having no idea where I was. The Tahitians, I'd discovered, have a nomadic way of dealing with sleeping arrangements. When they want some shut-eye, they simply pitch a light mattress down on the floor wherever they happen to be. If they don't feel like bunking down in the bedroom and would rather sleep in the kitchen or the living room, then that is where they go. The entire house is flung open to the cooling night breezes—there are no closed windows or doors (and virtually zero crime). It's like camping, but indoors.

Getting up quietly so I didn't disturb Van Bastolaer's wife, Yvanne, and their eight-year-old daughter, Rainia, laid out on the living-room floor, I tiptoed to the garage where Miller and Hornbaker were attempting to shoehorn their gear into the car. Things were tight. Photographing giant waves requires all kinds of specialized housings and fittings and waterproof rigs, and none of it is small or light. The job is a jamboree of complications, a never-ending list of highly specialized stuff to buy, maintain, haul around, wield, master, and in general worry about. Rearranging the cases for a fourth time, Miller stood back and shook his head. "We're grown men dragging hundreds of pounds of equipment all over the world," he said. "Yeah," Hornbaker said. "But I'd rather over-pack than have to hacksaw a bolt onto something in the middle of the night because I don't have the one piece of equipment I need."

During the first twenty years of his career, Hornbaker never stayed for more than three months in any one locale. Instead he embarked on decades of perpetual motion, roving between the northern and southern hemispheres in pursuit of the ultimate waves. Miller was equally peripatetic; one time he had spent almost three days traveling from San Diego to a remote Indonesian island for a shoot, got the images he needed, and went home, which took even longer. Five days later he was asked to return to Indonesia for another assignment. Both men's passports were gridlocked with visas and stamps and customs carnets. Like most top surf photographers, they had logged plenty of time in Tahiti.

Even though it was only five-thirty a.m., and this was just a prep day, the swell still miles offshore, Hamilton and Van Bastolaer were already long gone. Teahupoo was located forty miles away, on the southern end of the island in the less populated region known as Tahiti Iti (Small Tahiti). Once Miller and Hornbaker's gear was crammed in, with a microwave-oven-size space carved out for me in the backseat, we set off in that direction too.

As the light came up, first as a faint wash and then fully illuminating the island's kaleidoscope colors, I began to see the beauty of the place. Tahiti was a riot of flowers and plants and trees bursting out of every available patch of land, birds of paradise squeezing in next to banana trees and orchids; royal palms bursting with coconuts vying for space with avocado trees, red torch ginger, and flamingo flowers. Dogs ran onto the road, which wound past sapphire and emerald lagoons straight from Gauguin paintings. Knife-edged volcanic peaks towered above, so heavily draped in dark green rain forest that they almost looked black.

The island was shaped like a wonky figure eight, with the top circle about three times larger than the bottom one. We traced the edge of Tahiti Nui (Big Tahiti), past tiny stores with French signs and roadside stands selling monoi oil and papayas. Along the way we stopped for baguettes. We passed Taravao, the isthmus that links the two parts of the island, and continued on to Tahiti Iti. Our destination was literally at the end of the road, marked with a painted stone that said "Teahupoo: Mile 0."

The pavement winked out in front of a narrow bridge that spanned a freshwater river streaming down from the flanks of Taiarapu, an extinct volcano. The river ended in a crescent of black sand beach. "People live on the other side of the bridge, but there's no electricity or anything," Miller said. "It's like stepping into another century." We turned into the last house at Mile 0, a white, two-story place known as Mommy and Poppy's. It was a private home that morphed into a big-wave staging ground when conditions warranted, hard at the water's edge and ideally situated to get out to Teahupoo's break, about a mile offshore.

It was barely eight a.m., but the yard buzzed with action. There were men and Jet Skis and surfboards everywhere, with roosters scuttling among them. Surf stickers, posters, and autographs plastered the wooden walls of the house and the garage. Mommy, a pocket-size Asian lady in her sixties, emerged from the kitchen in a red apron and baseball cap and placed a skillet of eggs and sausage on a long outdoor table. Several guys in their early twenties sat there, gathered around a laptop that was playing a surf movie, and they reached for the food without breaking their gaze. Commenting on the film, they spoke in their tribal dialect:

"That's radical! That G-Land?"

"Ahhh, fuck that was heavy. He was so far inside."

"I was there, that day on Pipe. People were frothing."

"Brah, I gotta get back to Indo."

I noticed Hamilton across the yard, puttying a fin onto a surfboard. His movements were brisk and efficient, his arm muscles flexing as he worked. It was an improbable display of vigor from someone running on two hours of sleep. "I have a lot of energy," he said, when I mentioned this. "Especially when it comes to this stuff. I almost don't need to sleep." Every big-wave rider I'd spoken to had stressed the impossibility of getting a good night's rest before a large swell. Hamilton referred to this tossing and turning as "doing the mahi-mahi flop. Full pan-fried mahi. Up every hour, looking at the alarm clock."

Behind Hamilton, four Jet Skis sat on trailers at the top of a launch ramp. Van Bastolaer leaned over one of them, suctioning fuel with a length of rubber hose. The smell of gas wafted through the air. "Hey, keep those Frenchmen with their cigarettes away from here!" he yelled, gesturing at a group of photographers who had gathered in a corner of the yard to smoke. Then he laughed, flashing a set of brilliant white teeth. At first glance, you wouldn't look at Van Bastolaer and think "elite athlete." Where Hamilton and many of the others had hard lines and sharply defined edges, Van Bastolaer had rounded corners. Even his hair was curly. In the big-wave pantheon there were plenty of poker faces and end-zone stares, but Raimana was the kind of man who smiled with his whole body, his brown eyes radiating a deep joy.

Next to Van Bastolaer, a mechanic in a Red Bull T-shirt worked on the engine of a Jet Ski emblazoned with Red Bull logos. Beside him stood a surfer wearing a Red Bull baseball cap. This was Jamie Sterling, a compact, hard-charging twenty-six-year-old from Oahu. He and Hamilton struck up a conversation, and other riders drifted over to join in. There was an aura of nervous excitement, the adrenaline already starting to flow. As Hamilton continued to fine-tune his board, one of the younger surfers announced that he planned to wear a thin wetsuit under his flotation vest—not for warmth but to add another layer between his skin and the reef. "And if you have enough buoyancy," he reasoned, "you won't be driven down as hard."

Hamilton looked up from his work. The force that this wave unloaded made the idea of adding an extra millimeter of neoprene for safety seem absurd, like hoping an umbrella might cushion the impact of a falling anvil. The flotation vests were another story; all of the riders wore them now, and some men wore two. Undoubtedly they had saved many lives. But this practice had begun at Jaws, in the sixty-foot water where a fallen surfer might never make it back to daylight. The reef at Teahupoo, on the other hand, lay only three feet below the surface.

"I'm not as worried about flotation here," Hamilton said.

"Yeah," Sterling agreed. "It's not like Jaws."

Hamilton nodded, reaching for a screwdriver. "Jaws is all about the hold-down," he said. "Teahupoo is all about the bounce."

Sometime during the night the waves arrived. By the time Hamilton and Van Bastolaer left the house at four a.m., surf boomed against the breakwalls, and when they drove up to Mommy and Poppy's, they saw water washing through the yard. At dawn, Miller, Hornbaker, and I met our boat captain, a tuna fisherman named Eric Labaste. In the marina where we gathered, heavy surges made it hard to load the camera gear. The morning was clear and sunny, with a riffling breeze and a shape-shifting batch of clouds. It looked like a carbon copy of the previous day, except for one thing: this was a completely different ocean. Where the swell hit the barrier reef, a few miles out on the horizon, a thick band of white spray pulsed and flared like a ghostly runaway fire.

We were sharing our boat with three French photographers, all of whom Hornbaker and Miller knew well. On any big swell the shooters were key to the enterprise; as with the proverbial tree falling in the forest, if you ripped down the face of a hundred-foot wave and there was nobody there to take a picture, did you really surf it? There were many responses to that question, of course, but no one really wanted to answer it. In North America alone the surf industry is a $7.5-billion-per-year business, fueled by aspiration. Films, posters, magazines, screensavers— images of all kinds—are the currency of the realm, with giant waves in especially hot demand. So like the riders who lived separate lives until the weather map funneled them to the same destination, to Tahiti or Hawaii

or Australia or South Africa or wherever the next wave frontier material-
ized, the photographers reunited at the scene of the swells. Pushing off
from the concrete pier, Labaste took a last sip of his coffee and steered us
toward Teahupoo.

I heard it before I saw it, the exploding curtain of glass that ham-
mered onto the reef, the lip of a thirty-foot barrel hitting the earth like a
liquid apocalypse. From a visual standpoint, Teahupoo was a looker. Rich
lapis, deep emerald, pale aquamarine—its waters were the color of jew-
els, and its heavy white crest glittered in the sun. But even though the
wave was gorgeous, it had the personality of a buzz saw. As Teahupoo
reared up it drained the water from the reef, turning the impact zone—
a lagoon that was mercilessly shallow to begin with—into a barely covered expanse of sharp coral, spiky sea urchins, and volcanic rock. This
happened in seconds, in an area maybe three hundred feet long. I stared. I
had never seen a wave behave like this one. "Yeah, it's different," Miller
said, seeing my stunned expression. "Kind of like a shotgun unloading."

Though I could barely tear my eyes from the wave, I forced myself
to pull back and take in my surroundings. It was bizarre, really, how close
you could get to the ferocity. Because Teahupoo is created by a swell hitting a protruding knuckle on the barrier reef, there is—theoretically—
a safe channel right next to it where the water is deeper. Our boat and a
handful of others sat on the shoulder, so near to the edge that when a
surfer kicked out of a ride, he had to watch where he landed. Hamilton
had once torn his knee apart here when he exited a wave, trying to avoid
ramming into one observer's outboard motor.

Even in this so-called safe zone, however, the most experienced
boat captains stayed on their toes. They knew that the channel wasn't a
permanent fixture. It could suddenly vanish if the swell direction shifted
slightly, or if an especially huge set came shrieking in. Over the years
boats had been hit by the wave, flipped, and destroyed. And one time, Van
Bastolaer had been deep in Teahupoo's barrel and seen a hulking black
object whiz by only inches above his head; it was a Jet Ski that had been

catapulted over the falls when its driver ventured a little too close to the edge.

A fine mist hung in the gold morning light and when the wave broke, spray showered over us. Wind was minimal, giving the water the greasy, glassy texture that surfers love. "It's gonna be a show of power out there today," Hornbaker said, hefting the waterproof housing for a Super 16 movie camera onto his shoulder. Directly in front of us, as though engaged in some avant-garde stage performance, a Brazilian surfer dropped onto a wave. He wobbled in the barrel for a few seconds before being pitched into the air backward; the effect was of a bowling pin blown off a balance beam by a fire hose. We saw his board rocket into the sky, and the flash of a leg that looked like it was bent in the wrong direction. "There's a scenario," Miller said. "A bad one."

As the wave hit with grenade percussion, the surfer disappeared into the maw. His partner darted by for the rescue, and as he passed you could see in his face that Teahupoo's impact zone was the last place he wanted to go. Steering into the whitewater, he looked frantically for the surfer's head to pop up, but there was no sign. The driver circled, still searching, but the next wave was already bearing down. He was out of time. He was also out of luck: when he hit the throttle to speed away, his engine quit. "He's cavitating!" someone yelled from a boat. Jet Skis were notorious for this, for stalling in the churning foam, their motors grasping for traction from the water only to end up stuttering on air. Now, instead of a rescue vehicle, the driver was out there with a thousand-pound problem. Lacking other options, he dove into the whitewater.

Noticing the trauma, several teams rushed to the edge of the impact zone, ready to help. One of them corralled the abandoned Ski, while another managed to get the driver onto their rescue sled. As a third wave reared up, the surfer's head was spotted at the far side of the lagoon. He'd traveled more than five hundred yards underwater, shot like a cannonball across the reef. Someone plucked him from the water, sparing him further beating.

Five minutes passed, and then ten, and still the Jet Skis clustered at the bottom of the lagoon. In the channel, people speculated about snapped necks and missing limbs. Then the surfer appeared, splayed out on a rescue sled, grinning and waving as if he were in a ticker-tape parade. Blood dripped from his elbows. As he went by he flashed us a *shaka*, the Hawaiian hand sign for "things couldn't be better." "Oh, God is merciful," Hornbaker said. Miller lowered his camera and stared in disbelief: "He doesn't have any idea how fucking lucky he is."

Hamilton drove up to us on a red Jet Ski. From every boat, photographers trained their lenses on him. He wore a white, long-sleeved rash guard over an armored flotation vest, giving a kind of Ninja Turtle effect, and knee-length neoprene shorts. "Tickets on the fifty-yard line," he said, smiling. For Hamilton, this really was a day when things couldn't be better. His family and friends knew it well: the more waves Hamilton rode, and the greater the degree of difficulty of those waves, the happier and easier to be around he became. "If I scare myself once every day, I'm a better person," he had said. "It helps to have that little jolt of perspective that life's fragile." Few places made that point more clearly than Teahupoo. Out of respect for his host, however, Hamilton had begun the day driving, not surfing. He had just towed Van Bastolaer into an enormous wave, and now they were headed back out for the sequel.

Weaving between the vessels in the channel, a dozen tow teams motored back to the takeoff zone, which was known as the lineup even though nothing out there was orderly as a line. The area was so called because it provided visual alignment with a landmark on shore (in this case, a notch between the steep volcanic peaks), a cue the surfers could use to position themselves correctly. It was important local knowledge, a precise sense of where the wave would break.

For this and other reasons, eagle-eye vision was a critical part of the game. To have any hope of catching one of these stampeding giants, the riders had to spot it from afar. This wasn't an easy skill to acquire. When it first appeared on the horizon, a promising wave was merely a subtle shadow in the water, like a blurry strip of corduroy. As the energy neared the break, the water would rise into a lump. Some lumps, meanwhile, were lumpier than others, and those were the ones that everyone wanted. Often a half-dozen teams would end up gunning for the same wave, though only one man could ride it; to have multiple surfers tearing around at forty miles per hour on a giant face was undesirable and perilous. To determine who would get the wave a furious game of chicken ensued, played through a scrim of considerations that included who was best positioned, who let go of the tow rope first (placing himself deepest into the belly of the wave), how many waves each rider had (or had not) already caught, whose driver was most aggressive, who was higher up the big-wave food chain—in a few quick seconds all these things came into play. Tow surfing was not a sport for the timid or the excessively polite.

"This is a hell-raising group," Miller said, surveying the teams. "There's Garrett McNamara. G-Mac. I saw him get his leg sashimied here. Whole thigh ripped open, right to the knee. They called that day Bloody Sunday." McNamara, forty-two, was a highly skilled surfer with a wild streak the size of Interstate 10. His combination of talent and audaciousness drove him to do things that few others would attempt. Shortly before coming to Tahiti, for instance, McNamara had surfed the wave kicked up by a calving glacier in Alaska, dodging falling hunks of ice the size of city blocks. On another occasion, for a promotional video clip, he let an eight-inch-long centipede crawl out of his mouth.

McNamara's Don't Try This at Home persona also—quite regularly—led him to endure wipeouts that not everybody could survive. I watched him as he drove by on a camouflage-painted Jet Ski, wearing a camouflage-patterned rash guard and a black baseball cap. Though he was almost always smiling, there was a dark intensity to McNamara's presence. His hair was close-cropped and jet-black, his eyes were a color deeper than brown. Like many of the best riders, he grew up on Oahu's north shore and was forced to fight his way into that brutal surf fraternity. But his tough Hawaiian neighborhood of Waialua looked like a penthouse at the Four Seasons compared to his tow partner Koby Abberton's hometown.

Abberton, twenty-eight, came from Maroubra Beach, a suburb of Sydney, Australia. Maroubra was a feisty slice of coastline known for its challenging surf, vast sewage-treatment plant, maximum security jail, and dense population of heroin dealers and addicts, which included Abberton's mother and her boyfriend, a bank robber. It was a snake-eyed roll of the dice, Abberton's youth, so filled with violence and obstacles and, ultimately, surf salvation that even as he cruised through the channel at Teahupoo, his life story was being made into a documentary narrated by Russell Crowe. When Abberton was fourteen, he and his older brothers Sunny, twenty-one, and Jai, nineteen, had started a surf gang known as the Bra Boys. (A double entendre: *Bra* is short for Maroubra and surf-speak for "brother.") The gang, now four hundred strong, gained notoriety in 2003 when Jai Abberton was charged with the shooting murder of a Sydney man (he was later acquitted on the grounds of self-defense); Koby was accused of helping him dispose of the body (a charge for which he received a nine-month suspended sentence).

"Watch Koby's wave," Miller advised. "He won't even be on one unless it's totally insane. Bust the wall down or nothing." Sitting behind McNamara on the Ski, Abberton—who bore a striking resemblance to the actor Mark Wahlberg—looked pretty mellow for the moment, an impression that was undercut by his heavily tattooed neck.

As the morning progressed Teahupoo pumped out one snarling wave after another, but the swell had a distinct rhythm. It pulsed lightly and then heavily and sometimes convulsively. An extra-powerful set of waves would rush in, only to be followed by a relative lull until the next blast of energy arrived. "Every wave's different on the same day," Hamilton had explained, describing how minute changes in swell direction, wind, and interval—the number of seconds that passes between two waves—add up to endless variations. "It's never the same mountain."

There had already been one wave, ridden by a twenty-four-year-old surfer from Maui named Ian Walsh, that was freakishly bigger than the rest. This wave just had more of everything: more height, more girth, more foam and chop, more lunacy. It was as though Teahupoo had hated the taste of this one and spat it out in disgust. Collectively, people gasped. Walsh's wave was more like a wave and a half, and he knew it, throwing his head back in joy as he made a clean exit, and then bowing it in relief.

The wind came up, whipping spray and adding a toothiness to the water. Hamilton took a wave; immediately I could see the distinction between him and the others. Many times this morning I'd seen surfers straining to hold their own in the barrel. The wave sucked so much water up its face that unless the rider had the proper reserves of momentum and mass, they were quickly outmuscled. With his power and size, Hamilton didn't fly across the wave so much as he carved a trench through it. But his most startling move was to remain in the barrel for several beats longer than anyone else. Instead of racing ahead of the falling lip, he toyed with it, exiting at the last possible second as though stepping over a building's threshold the instant it collapsed. When he kicked out of the wave in a signature 360-degree flip, the channel erupted in cheers.

During the afternoon conditions got rougher. Clouds snaked around the base of the sheer peaks behind us, and the continued pounding of the waves churned the water so much that the ocean changed color from a clear azure blue to a muddy, foamy green. A log the size of a telephone pole floated into the channel. Before it could inflict any damage, the great Tahitian surfer Poto zoomed in on a Jet Ski and removed it. Although he wasn't riding on this day, Poto (whose given name is Vetea David) was royalty in these waters, Tahiti's first pro on the world-cup circuit. In person, Poto was a guy you'd look at twice. His movie-star features were toughened by a pugilist edge; think Polynesian Ken doll crossed with South American mobster. The image was completed by the stunning woman in a snip of a bikini who sat behind him on his Jet Ski, glossy black hair cascading down her back.

More rides, more triumphs, more wipeouts. "Bye-bye!" one of the French photographers said in his charming accent, as a surfer's legs buckled beneath him. Watching him bounce down the face of the wave, I could only wonder what was going through his head at that moment, what desperate prayers he was uttering. On the boat beside us one rider had stripped down to a loincloth, his upper body and hips raked with gouges and covered in blood. His board, broken into pieces, lay on the stern.

After a while Teahupoo spat out another superwave, caught by Shane Dorian, thirty-five, a veteran surfer from Hawaii's Big Island. As it reared up, everyone in the channel leaned back reflexively. Though he was one of the strongest riders around, Dorian seemed caught off guard by the behemoth and teetered on the brink of a fall. He managed to stay upright until the end bowl, the wave's exit, when he was blown out on his back. It was a near miss of a high order. "He was close to disaster on that one," Miller said. Dorian agreed: "I thought I was too. I was correcting the entire time."

Hamilton rode more waves, and so did Van Bastolaer, including one that he ended up sharing with Garrett McNamara. McNamara, however, had gone kamikaze, dropping in so deep he was doomed from the start. He bit the dust spectacularly. "I went toward the reef at like, 100 mph," he told *Surfing Magazine* later. " . . . I'm talking to God going, 'Please, please don't make this one too bad.' " McNamara escaped in one piece but not before taking "about ten" waves on the head due to his position in the center of the maelstrom. When Abberton was finally able to reach him, McNamara recounted, "I checked my cuts, hugged Koby, and went 'Thank you! That was insane! I feel alive!' " Later he would find shards of coral embedded in his helmet.

The day was winding down. The surfers looked spent, the spectators were thoroughly sunburned, the photographers rifled the coolers for cans of Hinano, the local beer. Only Labaste, still scanning the horizon, vigilant, and Teahupoo itself, continuing to grind out giant waves, maintained their energy. By sunset just about everyone had headed back to shore.

Mommy and Poppy's yard had a celebratory air, the riders high on relief and testosterone. They milled around attending to their gear, rinsing it, packing it, winching Jet Skis back onto trailers, drinking beer, and asking the photographers to show them digital images of their rides. When you're actually on a giant wave, they told me, you don't get the full measure of the animal. The experience is more like a collage of sensory impressions. There may be a flash of white spray, a sudden jolt, a feeling of energy surging beneath your feet, the suspension of time so that ten seconds stretch like taffy across a violent blue universe. Inside the barrel, a place that surfers regard with reverence, light and water and motion add up to something transcendent. It's an exquisite suspension of all things mundane, in which nothing matters but living in that particular instant. Some people spend thirty years meditating to capture this feeling. Others ingest psychedelic drugs. For big-wave surfers, a brief ride on a mountain of water does the trick.

I sat at the table listening to a group of them trade stories about the day, all of their tension and pre-swell fears burned off in the waves. Across from me Miller was talking to a Tahitian surfer named Teiva Joyeux. I'd met Joyeux, thirty-one, the previous day and been struck by his quiet, elegant manner. That, and his tattoos. They were done in the traditional Polynesian style and covered much physical real estate. Joyeux had inscribed patterned bands, swooping curves, and sinuous animals onto his arms, legs, back, stomach, chest—he and his wife, Nina, had even tattooed wedding bands onto their fingers. It might not be a look you'd want to adopt if you worked at McKinsey, but on Joyeux it was natural and striking. I knew that for him, days like this one were bittersweet. On December 2, 2005, his younger brother Malik Joyeux, then twenty-five and one of Teahupoo's star performers, had died in a fall while surfing at Pipeline, in Oahu. The two brothers had been close, and the loss was shattering. And so while a major swell brought friends from around the globe to Tahiti, for Joyeux there would always be one rider who was missing.

The yard began to empty out. Some men had planes to catch later that evening—on to the next wave. Leaving the pipsqueak rental car to Miller and Hornbaker, I climbed into the backseat of Van Bastolaer's truck behind Hamilton, in the passenger seat. He was relaxed, wrung out in his favorite way, and his voice was shot from yelling all day over the roaring waves. Both men's eyes were so bloodshot that they were painful to look at. (For obvious reasons, tow surfing does not involve sunglasses.) As we pulled out of the driveway, a rider I didn't recognize dashed up to the driver's side window and thrust his arm into the truck, trying to shake Van Bastolaer's hand. He was a small guy with long, matted hair, and he was beside himself. "Thank you! Really, THANK YOU!" he said, and then paused, searching for words. "That was . . . REAL."

"Real," Van Bastolaer repeated with a laugh. "Yeah, brah. No faking over here."

We headed down the road, leaving Mile 0 and all its magic behind. A filmy night descended on the water, the sky settling into layers of apricot and rose that slowly deepened to a violet-tinted black. Hamilton, eating a tin of smoked almonds, offered them around. "No," Van Bastolaer said. "I want to kiss my wife, and I want the real taste." He turned to me in the backseat. "Happy wife, happy life!"

Hamilton looked at him. "You could wash your mouth out with soap and she'd still be able to taste the beer."

Tahiti's not much on artificial illumination; no streetlights or flashing signs or lit-up office buildings drown out the blaze of stars overhead. I could still hear the waves crashing nearby, but the near darkness gave me a chance to process the day's sensory overload.

Teahupoo, with its timeless power, brought to mind the age-old philosophical quest to distinguish between beauty and its twisted cousin, the sublime: for the merely pretty to graduate to the sublime, terror was required in the mix. "The Alps fill the mind with a kind of agreeable horror," wrote one seventeenth-century thinker, summing up the concept. And while humans were capable of creating the lovely, the dramatic, the sad, or the inspiring, only nature could produce the sublime. It was a concept both comforting and disturbing: there are many things out there more powerful than we are. No one was more aware of this than the men who'd ridden Teahupoo on this day (except, perhaps, the ones who had fallen on it).

"Everyone's going to have Post Big Wave Syndrome," Hamilton said in a hoarse croak. This was his name for the inevitable low that followed an endorphin high. The body had squandered all of its good drugs in a single binge. Now a resupply was required—and that could take weeks of dragging around, feeling excited by nothing. "Sometimes it doesn't hit for three or four days afterward," he said. "Before I knew what it was, it used to hammer me."

"Ah, brah," Van Bastolaer said, "we're gonna have another big swell here before New Year's. I have a feeling. I'll be calling you." He mimed a dialing motion and laughed. "You'll be back."

PENETRATING SO MANY SECRETS, WE CEASE TO
BELIEVE IN THE UNKNOWABLE.
BUT THERE IT SITS NEVERTHELESS, CALMLY LICKING ITS CHOPS.

H. L. Mencken

SCHRÖDINGER'S WAVE

KAHUKU, OAHU

The north shore of Oahu is a lovely, blustery place that attracts a smattering of tourists, residents looking to escape the rush of Waikiki, and just about every serious big-wave surfer on earth. If the wave kingdom has a Hollywood or a Mecca or a Harvard, they are all here. "They call it the seven-mile miracle," Hamilton told me. "It's the proving ground. If you're a surfer and you're coming up—and you're serious about doing anything—you gotta go to Oahu and show what you've got. It's unbelievable how much surf there is in such a small area. And every wave is a great wave." Along the north shore there were lefts, rights, point breaks, perfect tubes, giant walls; there were inner reefs and outer reefs and on the truly massive days, waves would break on the outer *outer* reefs, where Hamilton and Darrick Doerner had first experimented with tow surfing and where Jeff Hornbaker had once filmed in waves so savage that in their aftermath dead sea turtles floated to the surface.

On a postcard-perfect evening in mid-November I drove down the two-lane road that runs along those twisty seven miles, from Haleiwa with its sign declaring it the "World Surf Capital," and then on past Himalayas, Alligators, Waimea Bay, Log Cabins, Back Door, Pipeline, Sunset Beach, Backyards, Velzyland, and Phantoms—a chorus line of famous waves.

Along the way I dodged cyclists and skateboarders and jaywalkers running barefoot across the road to the beach, all of them carrying surfboards. Big-wave season was in high gear, but I hadn't come to watch people ride those waves. I was here to listen to people talk about them. At the northern tip of the island I turned into the Turtle Bay Resort.

In the lobby sunburned families booked snorkeling excursions, and honeymooning couples drank mai tais and nuzzled. Trade winds blew in from the ocean. Hawaiian Muzak enveloped the place like a cloud. And over in Ballroom One, 120 scientists attending the Tenth International Workshop on Wave Hindcasting and Forecasting and Coastal Hazard Symposium milled around at their icebreaker cocktail party, temporarily tripling the north shore's per capita IQ.

Every two years the world's most eminent wave scientists gather somewhere to exchange information, present papers, compare notes, and above all, to argue. It was a big-wave group every bit as elite as the one that had descended on Tahiti, except that here the wave action involved risks of another sort; how human interests might coexist with things like fifty-foot storm surges, ship-busting rogue waves, and extra-strength hurricanes.

Contrary to stereotype, these scientists were a diverse, healthy-looking crew. There was the usual raft of thick glasses and tufts of facial hair, but there were also a heartening number of women in the group, as well as a younger contingent in baggy shorts and flip-flops that wouldn't have looked out of place down the road at Pipeline. After the 2004 Indonesian tsunami and the inundation of New Orleans, and amid growing concerns about how drastically climate change was likely to affect the oceans, wave science had become a hot topic, and fresh energy was pouring in. The field had come a long way since World War II, when military planners, realizing that stealthy beach landings required accurate surf forecasts, were dismayed to discover that nothing of the sort existed. (For scientists, nothing guarantees job security more than working on something considered useful for war.)

The conference chairs, Don Resio and Val Swail, stood by the registration table and greeted attendees. I walked over to introduce myself; I'd spoken on the phone to Resio, a senior research scientist from the U.S. Army Corps of Engineers, and he had agreed to let me come to the conference. Even as a disembodied voice he was instantly likable, but in person Resio had the kind of natural magnetism politicians dream of. He was a tallish, jovial Virginian with a silver brush cut and a neat goatee. When he smiled, which was often, he revealed a set of perfect white teeth. "Ah, hello!" he said, shaking my hand. "Welcome, welcome. It's wonderful to have you here." Despite my gate-crashing status, Resio meant it. He had only one concern: that I wouldn't understand a word of the conference.

It was a valid worry. At this level the study of waves involved quantum mechanics, chaos theory, advanced calculus, vortex turbulence equations, and atomic physics. I was a little rusty on those things. "Don't worry," Resio's co-chair Val Swail said, with a wry smile. "When they start in with the equations it's over our heads too." Swail, an outdoorsy-looking Canadian with a shock of gray hair and a ruddy complexion, worked on the leading edge of climate research for that country's government. He looked at me sympathetically. "And no one understands Vladimir [Zakharov]. He uses five integrals. The rest of us use two."

As if to illustrate their point, blowups of scientific papers lined the ballroom. They had titles like "Spectral Wave Modeling of Swell Transformations in Indigenous Marshallese Navigation" and "Blended Global High Resolution Sea Surface Forcing Parameters for Numerical Ocean Modeling." I looked at a printout of one of them, thick with math. The posters ranged from graphic extravaganzas—the kind of thing you get as a bonus when you subscribe to *National Geographic*—to one halfhearted effort consisting of a black-and-white piece of letter-size paper stapled to the wall. There were charts that looked as if they'd been sprayed with buckshot, and graphs that resembled solar systems. As I browsed the presentations I overheard snatches of conversation. "I was puzzled a little bit myself about the high-frequency tail breaking," a Japanese scientist said.

"Naturally," an Italian scientist replied. "You contend there is a power law for wind-wave interaction," a stern-faced man said in a strong German accent. "I think that is a questionable assumption."

On the far side of the ballroom I came across a paper I could actually understand, a colorful affair illustrated with photographs of giant waves crashing down, titled "Prototyping Fine Resolution Operational Wave Forecasts for the Northwest Atlantic." The bottom line on this one, it seemed, was that we are bad at forecasting wave behavior in the most extreme storms. "Are storms becoming stronger?" I asked its author, Canadian scientist Bash Toulany, standing nearby. "That's a tricky question," he said. "It has to do with sea surface temperatures and—"

A cheerful voice over my shoulder cut in. "Oh, we're gonna get smacked. No doubt." I turned to the speaker, a happy-looking dark-haired man in glasses, in his early forties. His name was Dave Levinson, and he was a climatologist from the National Oceanic and Atmospheric Administration (NOAA).

"Smacked?" I said. "By climate change?"

Levinson nodded. "We have some major challenges coming up that are pretty heavy-duty. Do you need another beer?"

For all his humor and ability to communicate with a layman, Levinson was a dead-serious expert, and the conference chair on climate change. He had an excitable air. When he spoke, the words poured out of him at high speed, as if everything was so insanely fascinating there might not be enough time to get it all across. "I'm into storms," he told me. "I've always been into storms." From growing up in Chicago, Levinson had fond childhood memories of smiting midwestern blizzards that blanketed the region: "All the cars would be off the road and we'd go cross-country skiing."

Although global warming means less snow, there are many indications that we won't lack for terrible storms. According to Levinson, we were in for major changes—if not outright fire-and-brimstone disaster— when it came to ocean behavior. "You've got a couple things going on," he said. "You've got storm tracks shifting. You've got water levels rising." This was undeniable. The global average sea level rose approximately 6.7 inches in the twentieth century, and the rate is accelerating: a conservative estimate for the next hundred years would add another twelve inches to current levels; some scientists believe it will be more like six feet.

Melting ice contributes to higher sea levels, of course, as do warmer ocean temperatures, because water expands as it heats. While scientists are divided over whether a warmer ocean will result in more frequent storms, they do know that the strongest storms are intensifying. (Warmer ocean temperatures also mean more wind, and hurricane strength rises exponentially with wind speed.)

At the same time, in certain areas, the North Pacific and the Southern Ocean for instance, wave energy is increasing. These brawnier waves have the potential to cause damage wherever they hit, but they are extra-destructive at the coastlines, where they cause severe erosion, property damage, and death. During any given year the news reports numerous stories of waves sweeping shoreline observers off piers and promontories and beaches.

In places where sea ice or coral reefs have acted as natural wave barricades, that protection disappears if the ice melts or the reefs crumble (from exposure to stronger waves). Storm surges can then make deeper inroads, and the whole feedback loop continues. When you take in the interconnectedness of the entire system—and the fact that nine of the world's ten largest cities are located on low-lying coastal land—the warming oceans are one huge, nasty set of tumbling dominoes. "You hate to be dire," Levinson said, "but . . ." His voice trailed off.

A wave might seem to be a simple thing, but in fact it's the most complicated form in nature. Scientists even find it difficult to agree on a basic, all-around definition of what a wave is. Many, but not all, waves move a disturbance through a medium. That disturbance is usually, but not always, energy. A wave can store that energy or dissipate it. Paradoxically, it's both an object and a motion. When wave energy does move through a medium—water, for instance—the medium itself doesn't actually go anywhere. In other words, when a wave rises in the ocean and appears to race across the surface, that specific patch of water is not really advancing—the wave energy is. It's like cracking a whip. As energy passes through the ocean, it spins the water molecules in a roughly circular orbit, temporarily lifting them. Only when the wave is about to break, on a beach, say, do the water molecules shift locale—and even then just slightly—as they pitch forward onto the sand.

In order to exist, waves require a disturbing force and a restoring force. In the ocean that disturbing force is usually, but not always, the wind. (Earthquakes, underwater landslides, and the gravitational pull of the sun and the moon can also play this role.) The restoring force is usually, but not always, gravity. (In minuscule waves it can be the capillary action of the water itself.) All of this goes to explain why, if you're serious about trying to pin down a wave, you turn to equations rather than words. Because waves do all sorts of bizarre stuff.

There's the standing wave, in which energy moves in two opposing directions, and the Love wave, which travels only through solids. Gamma-ray waves—tiny, superenergetic electromagnetic waves—can kill living cells. Many scientists have named waves after themselves, resulting in such mouthfuls as the Tollmien-Schlichting wave. The X-wave (short for "extraordinary") is a slippery beast; it appears to zip around faster than light, theoretically allowing it to move backward in time. Then there are the mysterious gravitational waves that, according to the theory of general relativity, flex the surface of space-time. But we have to take Einstein's word for that, because nobody's ever encountered one.

Despite their differences, waves do share some traits. They're defined according to wavelength, the distance between two consecutive crests; and period, which represents the same measurement in time. Taken together, wavelength and period determine speed: longer means faster.

(Tsunamis, waves caused by sudden lurchings of the earth's crust, are the ocean's speed champions. Their wavelengths can be more than one hundred miles long, and they can travel faster than jets.) Longer-period ocean waves, anything over twelve seconds, are the ones sought after by big-wave surfers because they contain the most energy and thus create the largest faces when they break. Their power comes from the wind transferring its energy into the water over a stretch of miles (a distance technically known as a "fetch"), so the most formidable waves emerge in places like the North Atlantic, the North Pacific, and the Southern Ocean, where storm winds yowl across vast areas of sea, long fetches uninterrupted by land.

Another thing waves have in common is that despite science's efforts to dissect them, they defy total explanation. Reading through a basic oceanography text, I came across this sentence: "How wind causes water to form waves is easy to understand although many intricate details still lack a satisfactory theory." One French scientist put it to me bluntly: "People have been studying waves for so many years, and we're still struggling to understand how they work."

The conference room was large, sunny, and triangular, with walls of windows looking out at the ocean. It was an idyllic place to discuss waves, like sitting in the prow of a glass-hulled boat. After a continental breakfast buffet and a traditional Hawaiian blessing to kick things off, people settled at long tables with their laptops. The session, Coastal Waves I, was led by a scientist named Al Osborne. He was a tall, solidly built man with wavy gray hair, wearing a blue hooded sweatshirt, and he stood at the podium with the casual manner of someone with nothing to prove. I was curious about Osborne, a Texas-born physicist who had attracted much media attention for managing to create freak waves in a simulation tank. Now based in Italy at the University of Turin, he had spent his career studying nonlinear dynamics in water, addressing the question of why some waves rolled along in a fairly normal fashion and others suddenly morphed into monsters. Generating them in a controlled environment was a major step toward figuring this out.

Along with his colleague Miguel Onorato, a thirty-seven-year-old prodigy who was also attending the conference, Osborne had discovered that while freak waves did not play by traditional physics rules (straightforward linear theories proving that, in essence, one plus one equals two), they could be partially explained using quantum mechanics, the more exotic equations that describe atomic and subatomic behavior (nonlinear theories as to why, in chaotic environments, one plus one occasionally adds up to seventeen). Things become weird when examined through the quantum looking glass. Matter and energy can exist as both waves and particles, depending on conditions. Reality is revealed as a flexible construct, studded with parallel universes. It seemed like fascinating and twisted territory, and I made a mental note to seek out Osborne later.

Onorato got up to present the first paper, "Three and Four Wave Exact Resonance Interactions in the Flat Bottom Boussinesq Equations." This was not a kind and gentle start to the conference. From the title onward I found it flat-out incomprehensible, and judging by the looks on people's faces around the room, I wasn't alone. Onorato was a slim and striking Italian with the disheveled appearance that, for a scientist, serves as visual shorthand for eccentric brilliance. *I may not get around to washing my pants,* the look says, *because I am too busy splitting this atom.* As he clicked onto a slide of equations so dense it looked as though chickens had stepped in ink and scrambled across the screen, a Chinese man to my left let out a sharp exhalation.

The bottom line, I was beginning to understand, is that wave science is mind-meltingly complex because waves themselves are that way. Elegant equations that describe how waves move through water were established back in the nineteenth century and are still useful today. But they were based on the notion that waves behaved in simple and predictable ways. While that may be the case if you drop a stone into a quiet pond, in the ocean the opposite is true: it is a field of constant, seething interactions between waves and wind and gravity; if each wave represents a note, then the ocean is playing the most intricate symphony imaginable. Teasing apart this chaos and corralling it into a neat package of numbers is a daunting prospect, but of all the exceptional minds that might be able to move the ball forward, many were in this room.

Onorato wrapped up and asked for questions. Someone coughed nervously. There was a moment of awkward silence and then a deep, heavily accented voice boomed out from the front. The questioner, Russian scientist Vladimir Zakharov, went on for quite some time. They didn't come any smarter than Zakharov, a barrel-chested man with snowy hair whose appearance brought to mind a smaller, friendlier Boris Yeltsin. I'd heard Zakharov, sixty-eight, author of the Zakharov Equation, described as "the father of nonlinear wave mechanics." Of the fifty-three topics he had listed as his "research interests," tossed in somewhere near the middle was "Construction of new exact solutions to the Einstein equations." He and Onorato volleyed back and forth, speaking what seemed like their own private language, and then a third man joined in. This was Peter Janssen, a Dutch scientist from the European Centre for Medium-Range Weather Forecasts (ECMWF) in Reading, England. He was another titan, at the vanguard of wave research. Some of the world's fastest computers resided at the ECMWF, humming away on Janssen's latest initiative: a marine forecast that attempted to predict the appearance of rogue waves.

The presentations continued in a blur of wave theory while outside the real waves grew. Surfers streaked past, filling the windows. At the podium, a scientist discussed Wave Watch III, a mathematical model that simulated conditions in the global seas. Models are the linchpin of wave (and climate) science. Essentially, they're colossal computer programs that interpret millions of readings from satellites, ocean buoys, wind arrays, weather balloons, and other sources. All of these data are being fed into the models constantly. The result, hopefully, is an ongoing picture of sea states, wind conditions, pressure zones, ocean circulation—and the interactions among the four—that can be used to forecast future climate behavior. Serious elbow grease goes into creating a model. Scientists are always rejiggering them and fine-tuning them for greater accuracy. Models like Wave Watch III are critical tools, massive scientific initiatives. Anyone who is trying to do anything anywhere near the ocean relies on them. There is only one problem with models: they are often wrong.

Models, don't forget, said that rogue waves were impossible. They demonstrated why the Draupner oil rig's engineers didn't need to worry about an eighty-five-foot wave showing up. They assured naval architects that the *München* was unsinkable in any storm. Models underpredict and overpredict and strike out completely at regular intervals. "We've got very sophisticated wave models," one scientist told me. "They're trying to reproduce what's going on, and they've been stretched to the best performance you can get with the physics that's in them. And yet they're not reproducing the waves properly under certain conditions."

When you consider the overwhelming complexity of the ocean and the atmosphere, it's not hard to understand why. It seems futile to try to model things so immense and so inherently random. To science's deep frustration, nature regularly confounds our attempts to predict its next move. When I'd asked Don Resio whether he thought climate change would lead to stormier oceans and bigger waves, he shrugged dramatically. "We can't predict ten days out," he said. "What makes us think we can predict ten years into the future?" At the same time—because we need whatever intelligence we can gather about the natural world in order to live in it and build things in it and hope to understand it—there's nothing to do but try. So the sessions continued, and people leaned over their laptops and tried not to notice through the windows that playing in the waves looked like a whole lot more fun than writing equations about them.

Driving from another part of the island, I got wildly lost on my way back to Turtle Bay and ended up tiptoeing into the Climate Change session, midpresentation. The first words I heard were: "There is high uncertainty here." Given what the scientists had been saying, this sentence seemed to sum things up perfectly. The next talk was about whether the waves were getting surlier in northern latitudes (yes), and what this might mean for ship navigation (trouble); followed by a consideration of how climate will affect hurricane frequency (we're not sure) and storm surges (pass the sandbags); all of which provoked heated discussion. During one presentation Zakharov, wearing a scarlet Hawaiian shirt, stood up and unleashed such a torrent of protest that Resio cut in: "Is there a question in there, Vladimir?"

Later in the day, during a talk about storm surge behavior, Resio mentioned Hurricane Katrina. As a Southerner he took that storm personally, and when he spoke of its terrible impact, his voice tightened and his eyes became grave. "When there's a wave event at the coast we always underpredict," he said, then paused. "Katrina was a wakeup call. We don't have the science that we as a nation need."

At the break I went outside, where I ran into Dave Levinson and John Marra, another scientist. Marra, who lived on Oahu, had longish hair and an athletic build, and he revealed himself as a surfer by staring at the nearby waves the way a cat stares at birds. When Levinson introduced me and described my project, Marra had a question. "Those guys who want to launch their melons off a hundred-foot wave," he said, "are they mentally ill?"

"Does that mean you think it can't be done?" I asked.

"I don't know the phase speed of a hundred-foot wave," he said, turning serious in an instant and citing advanced math theory about breaking waves. "I'd have to actually calculate the celerity. I don't see why not, I guess—if you're moving fast enough. But is it human nature to want to *do* that?"

I defended the tow surfers' sanity for a few moments, then steered the subject to climate change. Like Levinson, Marra was not investing in coastal real estate anytime soon. His opinion was grim. "The polar zones are done," he said, with finality. "And there are going to be basic survival issues like no water. Ecosystem collapses. Food chain—"

"There's no question you're right, John," Levinson said, interrupting. "The water issue's going to be the worst."

"Then what's with all the arguing in there?" I said. "Why don't scientists agree about this stuff?"

"Well, it's natural variability versus human influence," Levinson said, explaining how science had to discern which changes could be chalked up to nature's regular cycles, and which were due to our monkeying around with the planet's chemistry: "The record is so short that the naysayers can point to uncertainty in the attribution of climate change." You couldn't pin any single event on climate change, or even a few years of wacky weather: you had to examine longer-term trends. Which, of course, takes time.

Marra jumped in. "Dave's point is that you're not going to be able to prove it until it's too late. It's the frog-in-the-pot analogy. He doesn't know he's cooked until it's too late and he *is* cooked."

Levinson nodded. "Uncertainty doesn't mean it's not happening."

"I really wouldn't want to be a weather forecaster right now," Marra added, "because we're entering into a time when, possibly, there is no normal anymore." He smiled, shaking off the gloom. "But that's all the more reason to surf! Go grab a couple of those total 'now' moments, because that's all there's gonna be anyway."

I caught up with Al Osborne at the conference luau. Tables had been arranged on a bluff, backed by a silky pink sunset. The surf was still hopping. When the moon came up, tiki torches were lit and the wave crests flared in the background. A Hawaiian band took the stage, the singer leaning into the microphone: "Now, some hula for you wave junkies!" Osborne was sitting next to Resio, and I went over to join their table.

Osborne was nursing a head cold and some terrific jet lag, but he was still game to talk about waves. Actually, the subject seemed to perk him up. "All physical phenomena are waves," he said, with a hint of Texas drawl. He jammed his hands in his pockets, leaned back in his chair, and nodded toward the sky. "The universe is constructed of waves. And let me tell you why. It's craziness!"

From the start of his career, Osborne had a knack for running into the strangest waves. Trained as a cosmic ray physicist, he left NASA's Apollo program in Houston to work in Exxon's oceanography group. It was the early 1970s, and oil exploration was heating up. To safely drill in the ocean, companies like Exxon desperately needed science expertise. Osborne had this—for space. "When I first went to work there [Exxon] I didn't know anything about [ocean] waves," he said. "I mean *anything*. And water waves are more complicated than electromagnetic waves because they're nonlinear."

But he was a quick study and a curious guy, and when Exxon sent him on a critical assignment in the Andaman Sea, Osborne was more concerned by the State Department's warning that the area was inhabited by "implacably hostile headhunters" than any professional shortcomings. His job was to find out why the *Discover 534*, the world's biggest drill ship at the time, was getting boxed around in what seemed like calm seas. "Nobody had ever made any measurements there," Osborne said. "Nobody knew." Using jury-rigged instruments and an unintentionally expensive trial-and-error approach, Osborne solved the mystery. To his shock, he found that two hundred yards below the ship, gargantuan underwater waves more than six hundred feet high and ten miles long were rumbling by at a speed of four knots. Nothing in the model explained this. "It was just stuff that shouldn't have been there," Osborne recalled. "So I learned about internal waves."

We now know that internal waves are basic features of the ocean, visible on photographs taken from space. Water densities vary in any patch of sea, sort of like a layered cocktail. When tidal forces drag one layer over another, internal waves are born. They play a critical role in ocean circulation (and thus, climate) and move nutrients throughout the water column. And they have an additional characteristic that fascinated Osborne: they're solitons, waves that behave like particles.

Solitons have long been causing wave scientists to scratch their heads in bewilderment. In 1834 a Scottish engineer named John Scott Russell happened to see an odd wave moving through a canal near Edinburgh. Describing it as a "rounded, smooth, and well-defined heap of water," he followed the wave on horseback for two miles as it cut through the canal like a shark's fin, rather than oscillating up and down like a normal wave. "It rolled forward with great velocity," he wrote, "without change of form or diminution of speed. How could this be?"

The vision of a singular, misbehaving lump of water blew Russell's analytical mind. The wave, which he called the "Great Wave of Translation," seemed to defy Newton's laws of hydrodynamics (which consider fluids to be continuous fields rather than a parade of discrete objects). The Great Wave would obsess Russell for the rest of his life, and when he failed to convince other scientists of what he had seen, it ruined his career. It was only when quantum physics came along seventy years later and explained the soliton—how one wave could behave independently from, and be unaffected by, the waves around it—that Russell was vindicated.

As it happens, solitons exist anywhere there is wave motion: in gases, in telephone signals, in the sky, even in our bodies. But in 1975 for Osborne to find them lurking in the ocean was such a scientific coup that he ended up on *The Tonight Show with Johnny Carson*.

"I couldn't stand prosperity," Osborne said wryly, recalling those years. "So I moved to Italy." At the University of Turin he continued to study waves, examining their quantum underpinnings. In 1999 a graph depicting the Draupner freak wave spiking out of the ocean stopped him in his tracks. It looked exactly like a soliton. Therefore, his mind sped ahead, it should be possible to create freak waves using the nonlinear Schrödinger equation (a famous breakthrough in quantum physics that described this kind of renegade wave behavior by electrons).

Sure enough, in the wave tank Osborne was able to dial in Schrö-dinger and make tiny freak waves jump out of the water. "After hundreds of years of everything about freak waves being based on anecdotal evidence," Osborne said, "suddenly there was a real physical dynamic." While freak waves are not solitons exactly—the two are more like second cousins—the point he made was important: when you veered off the linear path into the dark, nonlinear woods, you came closer to understanding the ocean at its most extreme. (Since then tsunamis have also been identified as soliton kin.)

The longer I listened the more it seemed that Osborne's discoveries supported the scientific adage that "the universe is not only stranger than we imagine, it is stranger than we *can* imagine." When talking about his work, Osborne used phrases like "black magic" and "utter miracles," which would seem to make for a pretty exciting day at the office. In light of this I had one last question, a fairly preposterous one. Earlier he had mentioned that in order to come into being, a rogue wave had to mug (my description) its neighbors. After attempting to explain this to me using the Benjamin-Feir instability theory, the Reimann Theta functions, and Fourier analysis, Osborne finally broke down and personified the waves like sock puppets. "It's like this rogue wave is hiding," he said, using his hands to demonstrate. "He's got his arms out, covering a lot of other waves. And when he gets ready, he sweeps in all their energy, stealing it from them and pulling it up under this one big single peak."

Since I am not a scientist, the image of rogue waves as clever oceanic criminals delighted me and stuck in my mind. But what made one wave a perpetrator of this energy theft, and another a victim? It was almost as if some kind of ghostly, esoteric intelligence were involved. "I don't think you can call it intelligence," Osborne said quickly. "Intelligence implies that you can plot and scheme. That requires a brain." He paused for a moment and grinned. "But they do kind of play these games, don't they?"

The Extreme Wave session came on the fourth day. Though I'd expected some heavy beach attrition by then, it was a packed house. The first presenter was Luigi Cavaleri, an animated Italian in his sixties, with fiery eyes and a mellifluous accent. Cavaleri's talk was a cautionary tale about an exceptional—and completely unforecast—storm that had walloped Venice in 1966, the worst deluge in that city's history. If such a thing were to happen today, Cavaleri wanted to know, would we be able to predict it?

It was impossible not to like Cavaleri, a whip-smart man in a no-nonsense plaid shirt, his sleeves rolled up, his caterpillar eyebrows jumping around on his face, his hands swimming through the air. "How many of you have seen the sea from below in a storm?" he asked. "It's a completely different picture." There was a murmur of polite laughter: *as if.* Almost every wave scientist I'd spoken to had confessed a preference for land, admitting to ravaging seasickness and a distaste for getting batted around at sea when the real work took place in front of a computer. Though Cavaleri's research had revealed that, yes, we would probably be able to predict the storm now, his gentle reminder that it wouldn't hurt for wave scientists to actually spend some time in the waves was what struck me the most about his talk.

In this regard the surfers had an advantage when it came to understanding the most extreme seas. Feeling a seventy-foot wave rising beneath your feet, hearing its turbine roar, pulling three G's on its face, and then dancing away from the blast—all of this, while not the type of fieldwork that'll win you a Nobel Prize, is at the very least an informative experience. I had noticed that any tow surfer worth his foot straps was also a closet meteorologist, able to translate buoy readings, spectral analyses, swell periods, wind directions, and bottom features into the waves that would likely result. Many times Hamilton had surprised me by holding forth about things like wave refraction and dispersion, and

Kalama had such a knack for interpreting storm data that he'd been nick-named Decimal Dave. Perhaps there was an extra incentive to fathom the waves when your life, as well as your paycheck, depended on it.

A sandy-haired, studious-looking man named Johannes Gemmrich followed Cavaleri with a presentation titled "Are 'Unexpected' Waves as Important as Rogue Waves?" Unexpected waves, he explained, were super-size normal waves that happened along, up to twice as big as the average. More common and less mysterious than purebred rogue waves—which could reach heights more than four times greater than the surrounding seas—they could be equally destructive. Gemmrich showed slides of a trail on Vancouver Island where unexpected waves (sometimes referred to by sailors as sneaker waves) regularly sucked hikers off the rocks to their deaths. "Unexpected waves are not rogue waves," Gemmrich said. "And not every rogue wave is unexpected." I wondered how important the distinction was when you were being swept away by one.

Peter Janssen was up next, and he unfolded his tall, wiry frame from his chair and strode to the podium. He had wild gray hair, a peppery beard, and a strict, professorial appearance that seemed intimidating until you noticed the sparkle in his eyes. His talk concerned the second-generation version of ECMWF's freak-wave warning system, soon to be launched. He stood at the screen with one hand in his pocket and the other gesturing to a mash of numbers, Greek letters, symbols, dots, slashes, and square-root signs. In the most rudimentary way I could follow him, because along with the very fast machinery that was whirring in Janssen's head, he had a gift for being able to translate arcane wave science into plain English—even though English was his fourth language.

The previous day we'd met for a poolside lunch to talk about the warning system. "How can you possibly predict a rogue wave?" I asked. It seemed like a contradiction. "I prefer the term *freak waves*," Janssen said. "Rogue waves—I'm always thinking of a herd of elephants." He laughed and took a swig of Longboard Island Lager.

In his precise Dutch accent Janssen explained that in certain conditions, freak waves became far more likely to occur. The trick was to forecast those conditions. Surprisingly, the main criterion was not huge seas (although that helped) but rather the shape of the wave spectrum—the measure of how wave energy was distributed in a given area. It came down to steepness. Steep waves were farther from equilibrium: less stable, more prone to pirating other waves' energy. If the spectrum was narrow and peaky—as if someone had taken some lovely, rolling waves and squeezed them hard in a vise—that was when, as Janssen put it, "you get a very high probability of extreme events."

Fast-growing storms tended to create steep waves, as did high winds that blew for a long time in the same direction the waves were traveling. There were also infamous freak wave haunts like the Agulhas Current off the southeastern coast of Africa, where fast, warm currents collided head-on into colder, opposing currents, creating an oceanic train wreck. Again, this steepened the waves and deepened the troughs between them.

The ECMWF's method of predicting freak-wave probability involved dicing the seas into forty-by-forty-kilometer squares, setting a baseline, feeding ocean and atmospheric readings into the model, and then sounding the alarm when conditions looked suspect in any of the squares. In theory this sounds like a simple enough thing to do; in practice it is diabolically hard. How do you check to make sure your models are on the right track? At any given time, the oceans are mostly empty and unsurveyed, no one reporting back any excessively steep wave sightings. "It's difficult to validate our theories," Janssen said. "I am hoping to find satellite instruments that will be able to monitor those extreme situations." He shrugged. "It might be that we are completely wrong."

I asked Janssen the question I'd asked many other scientists: Should we expect more aggressive waves due to climate change? Like all of them, he hesitated before answering. "Wellll, what we see at the moment, yesss," he said, drawing the words out carefully. During an earlier session, he noted, Russian scientist Sergey Gulev had presented a paper showing that wave steepness had increased sharply between 1970 and 2006.

"Climate change is not easy," Janssen added. "Because in the early days we had hardly any data." It's tough to say there are more giant waves now—or hurricanes or windstorms—if no one knows how many there were before. Yet at this point few scientists believe there is nothing to worry about. "I can tell you one thing with climate change," Janssen said. "I am quite sure that it is happening."

It was November 2007, the tail end of an odd and tempestuous year. Huge waves had pummeled Europe, South Africa, Indonesia, China, Taiwan, and Australia, generating headlines like "Residents Flee After Waves Batter Indonesia's Coastline," "Asian Beaches Reopen After Winds Trigger Huge Waves," and "Giant Waves Thrash Reunion Island."

"Ireland Braced for Giant Waves: Massive waves higher than houses are expected to lash Ireland's west coast this weekend," warned the *Edinburgh News*, prompting one online reader to complain: " 'Waves higher than houses'—what kind of warning is that? Is that great muckle skyscraping houses, or wee tottie tattie-pickers' houses? We need to be told!"

Globally there had also been brutal heat, extended droughts, raging floods, runaway wildfires, and swooning fluctuations in temperature. Europe had been scoured by wind and deluged with rain. So much Arctic ice had melted that the Northwest Passage, historically impassable, had opened. Hurricane Noel, the sixth and deadliest hurricane of the season, hit Canada's maritime provinces on November 4, a late-season tempest that announced itself with eighty-mile-per-hour winds and fifty-five-foot waves. The waves swept away piers, toppled boats, flipped cars, tore up hunks of pavement, washed out roads, and tossed large boulders far inland.

One of the year's most dramatic incidents had occurred only days before, on November 11, the first day of this conference. As if to remind the scientists why their work was important and what could go wrong in the waves, a storm near Russia's Black Sea had sunk four bulk carriers and split an oil tanker in two, causing a three-thousand-ton oil spill. Foundering in seventy-mile-per-hour winds and thirty-foot waves, the bulk carriers had also tipped seven thousand tons of sulfur into the drink. Another six cargo ships ran aground and more than forty vessels were evacuated from Port Kavkaz, 550 miles south of Moscow. Three sailors were confirmed dead, with fifteen missing. As rescue helicopters searched the waves for survivors, warnings of a second storm were issued. Interviewed for TV, a port official was beside himself. The cameras rolled as he lamented the loss of tourism in the area, the chemicals in the water, the oil-slicked seabirds, the madness of the storm. "The ships weren't meant for these kinds of waves," he said, wringing his hands. "They shouldn't have gone out."

A WAVE IS A COMMUNICATED AGITATION.

Jack London

KARMA, TIGER SHARKS, AND THE GOLDEN CARROT

PAIA, MAUI

Days of rain had given way to a breezy and clear Sunday morning, palms tossing, mauve clouds slinking away on the horizon. By seven-thirty a.m. Hookipa Beach Park was already crowded, the surf breaking large and steady. An offshore wind lifted the waves as they charged toward shore, holding them open, it seemed, just a little bit longer, letting them growl and spit and lunge for an extra second or two. They were double overheads, aligned in perfect formation. Instead of closing out suddenly in crashing clumps of whitewater, they peeled neatly from top to bottom. White feathers of spray streamed off their peaks. Truck after truck pulled into the dirt parking lot, bumpers hard at the sand, and when the surfers jumped out and stared at the ocean and took in the waves, the wind, the perfect day-ness of it all, they moved with the urgent energy of kids on Christmas morning, bee-lining for the tree. Boards were slung down from racks and hauled out of pickups and zipped out of cases and waxed and examined for dings and then tucked under their owners' arms as guys loped across the beach. It was as if a silent signal had gone out, a kind of dog whistle for surfers, and summoned them to the water.

Hamilton arrived in his black Ford 250 pickup, windows down, Pearl Jam providing the sound track. As he rolled through the parking lot, he flashed a couple *shakas* and yelled a few "howzits" to friends. There was nothing so obvious as a huge grin on his face, but if you spent time around Hamilton, you learned to sense his moods. His energy was the high-octane sort; along with the extra power, there was a heightened risk of detonation. When Hamilton was frustrated or upset, his whole presence signaled it. His eyes flashed a duller color than their usual sea green and hardened into a disconcerting stare, his movements tightened, his voice became lower and flatter, his muscles flexed as though spoiling for a fight. He was known for this aggression, and for its flip side too, for ease and generosity and humor. "Laird can be extremely everything," Dave Kalama had explained one time. "He can be extremely kind, extremely patient, extremely irritating, whatever. I mean he's not your standard person." You have to remember the context, Kalama added. "People who aren't used to being in situations like we are don't understand—that intensity is part of what helps you survive."

But today Hamilton was happy. The waves were here. Even though these were little canapés for a big-wave rider, they were tasty, and in time the main course would arrive. With Thanksgiving only a week away, the winter wave factory in the Aleutian Islands was open for business; any day the weather radar might glow with the pulsing magenta blob that meant a serious storm, spidering down from Alaska. Some of the most memorable swells in history had shown up along with the turkey and stuffing.

Hamilton drove to the far end of the beach, parking in the shadow of the sunny yellow lifeguard tower in front of a sign that said: "Warning: Strong Current. You Could Be Swept Away from Shore and Drowned." Next to it stood another sign: "Mai huli 'oe I kokua o ke kai!," Hawaiian for "Never turn your back on the ocean!" On top of it someone had plastered a sticker of a red circle with a slash through the middle, the universal symbol for "no more of whatever's inside the circle." In this case it was a drawing of a standup paddle surfer, atop a scrawled warning: "Stay out of the surf breaks!" Hamilton got out and unloaded his thirteen-foot standup board, propping it against the sign as he reached into the truck for his long Kevlar paddle.

Standup, as surfers refer to it, is a twist on the sport that Hamilton, Kalama, Lickle, and others had taken up several years ago to challenge themselves on the days when sixty-foot waves weren't available. While the entire surf industry was gravitating toward shorter, quicker, thinner boards for doing flashy aerials on smaller faces, Hamilton was showing up on enormous planks that ranged from ten feet to more than sixteen feet long, and carrying what looked like an eight-foot outrigger canoe paddle. At first no one could quite figure out what he was doing, but once they did the sport exploded in popularity.

"We call it 'the ancient sport we've never seen but we know existed,'" Hamilton said, hoisting the forty-pound board on his shoulder. "It looks simple, but it's not." Unlike in regular surfing, the standup rider remains on his feet at all times. There is no steering around on the stomach, no hands in the water. Instead, the paddle is the means of propulsion. To catch a wave the trick is not only to stay upright (about as easy as balancing barefoot on a basketball) but to maneuver the hulking board into the right position on a dime (about as easy as performing dressage moves in a school bus).

As the sticker made clear, regular paddle surfers viewed standup surfers with varying degrees of annoyance, mostly near the high end of the scale. One complaint was that when a standup surfer fell, his outsize board became a runaway wrecking ball. Every surfer I met had been nailed numerous times by his, or someone else's, surfboard, and the aftermath was never pretty. Exhibit A was Hamilton, his body a wound map of punctures and divots. "I got my Pe'ahi gun through my face last winter," he had told me, describing how the surfboard's sharp tip had "exploded" the inside of his mouth. "It was like a fourteen-foot, sixty-pound spear gun right behind my teeth, through the gums," he added. "If it had hit me in the temple, it would've been game over." Then there was the cross-shaped scar on his left thigh, a gift one Christmas Day on Kauai when someone's loose board hit "like a pick ax to my femur." Or the time a surfboard T-boned him in the forehead at Pipeline: "134 stitches to my frontal lobe." I knew one surfer who'd had his eyeball split open, and another who'd been pierced in the C-2 vertebra, an encounter that left him temporarily paralyzed. When you considered the damage even a snappy six-foot swallowtail could inflict, a thirteen-foot missile knifing through the surf became a justifiably terrifying notion.

The other reason for resentment at a crowded surf break—and likely the more significant one—was simple: standup riders caught all the waves. While the other surfers, sitting on their boards, eyeballed the incoming sets from what Hamilton referred to as a "worm's-eye view," standup riders could see clear to the horizon. They identified the best waves early, and then used their paddles to accelerate past any other takers. I sat on a picnic table and watched Hamilton slicing his way toward the Point, a break about three hundred yards offshore, to demonstrate the practice.

A moment later Dave Kalama arrived in his white pickup, pulling up next to Hamilton's. His cousin Ekolu Kalama, thirty-one, a tall, regal-looking Hawaiian, was in the passenger seat. They got out, and took down their standup boards. Waving a quick hello as they crossed the sand, both Kalamas disappeared into the surf.

I spotted Hamilton through my binoculars, stroking past dozens of bobbing heads in the lineup, paddling farther out and snagging waves before anyone else had even noticed them. He was doing laps in perpetual motion—paddle out, surf in, paddle back out—as though trying to set a record for waves surfed within a period of time. Kalama soon joined him. For most surfers, this was the day they had been waiting for; for Hamilton and Kalama it was only training for that day, a chance to log some miles and experiment with technique and press the boundaries of their endurance.

As the morning progressed, the waves strengthened, kicking up more spray and breaking with a heavier, rounder sound. The air vibrated with energy. It smelled of water and salt and earth, with a slight fish tang. I watched as people tried, and mostly failed, to catch waves, their bodies gobbled up by the whitewater. A lifeguard shot to the beach on a Jet Ski and deposited a jangled-looking guy with a broken board. The surfer tipped off the Ski and staggered onto the sand, while the lifeguard pivoted a quick U-turn to avoid getting caught broadside. He gunned the engine, racing full tilt up the face of the incoming wave, barely clearing the crest as it broke. The Jet Ski took vertical air, almost flipping him over before landing precariously on the backside. Standing next to me, taking it all in, pulling hard on a cigarette, a snaggle-haired rider with a furrowed brow seemed to reconsider his plans. At the far end of the break I saw Hamilton and Kalama shooting down a wave while having what looked like a conversation; they faced each other because Kalama was riding his board backward.

Just down the beach Brett Lickle exited the water, so I walked over to talk to him. Few riders had spent as much time along this stretch of ocean. From Spreckelsville, an unruly, windswept area six miles down the coast, to the breaks here at Hookipa, to Jaws, five miles up the road, Lickle knew Maui's entire north shore with the kind of familiarity that enables a person to walk around his bedroom in the dark without knocking things over. I wanted to ask him why so few surfers were making their rides. "Oh, that's typical," he said. "Usually you'll see about six people doing all the surfing. The others are pulling into closeouts, with no idea of where the wave breaks or how it breaks. I call it reckless abandonment."

A wave's first appearance was subtle, more like a jot note on Nature's to-do list than a tangible thing. "A lot of the time you're reading shadows," Lickle said. "You know there's energy, but you don't know exactly where. But then you see that a certain shadow is deeper to the right. You've really got to be in the pole position if you want to catch one."

Like Hamilton, Lickle's passion for waves inspired him to constantly invent new ways to ride them. The garage at his house in Haiku was a dense maze of tools and parts and old boards, a mad jumble of possibilities. Surfboards, standup boards, tow boards, snowboards, skateboards, wakeboards, windsurfing boards, kitesurfing boards, even an ungainly contraption called a surf bike—Lickle had them all, in profusion. It is a matter of which toy was right for the moment and how the gear could be adapted for extra fun or difficulty. "You have to understand," he said, gesturing at Hookipa. "For us this is like trying to get a buzz on the kiddie roller coaster. You have to stand up or be upside down or something." He shook the water out of his hair and reached for his towel. "You have to tweak the variables or you'll die of boredom."

Fun, in other words, required a hard squirt of adrenaline—or it wasn't really fun. Lickle had described a game they used to play called Sky Pilot, a variation on tow surfing where the driver would slingshot the rider *up* the face of the wave, so it served as a moving launch ramp. "You would jump the wave," he explained, "go as high as you possibly could, do as many rotations as you could, and then land out back on the flats." He laughed. "We'd go thirty feet in the air." What eventually happened to Lickle while playing Sky Pilot goes a long way toward explaining why today it is no longer on the menu: "I landed one time and everything just gave; I literally brought my heels to my butt. It blew out all the ligaments in both knees. Another time the lip of the wave hit one knee and buckled it backward, and then I hyperextended the other in the opposite direction."

Throughout his career Lickle had endured the standard amount of trauma, horrifying to the average person, par for a big-wave surfer. He'd been over the falls at Jaws, held down beyond all reason in the pre-flotation-vest days, rammed in the groin by a Jet Ski. He'd had his share of bone breaks, contusions, and near misses, and at forty-seven, with a wife, Shannon, and two daughters, McKenna and Skylar, he still remained in the center of the action even on the craziest days. The key to this longevity, he believed, was knowing when *not* to go out. "There are days when I think, 'No. Scrap me today.' If you don't feel it, you don't want to push it."

Fearlessness might seem like a basic requirement for big-wave surfing, but in fact the opposite is true. "Just to sit in the channel and listen to Jaws unload is enough to scare you out of the water," Lickle said. "If you can see something like that and not be scared, you gotta have something missing. Or you're terminally ill. You got something you don't care about." I knew Hamilton felt the same way and even took it a step further. "Fearlessness is ignorance, and it's lack of respect," he said when the subject came up. "Fear is powerful. You get a lot of energy from fear. Without fear, humans wouldn't have survived. Maybe I'm the *most* scared."

But if fear was healthy, panic was dangerous. A famous saying in big-wave surfing was: "Everything's okay until it isn't." When things go wrong on a seventy-foot wave, Lickle said, "you've got issues." He chuckled knowingly, nodding at the skyline as though he expected trouble to come from that direction. "The key is not to freak out. You freak out, you expend your resources." Personally, I found it hard to imagine relaxing in the middle of an underwater bomb blast, but apparently this was the trick to survival. If you kept your cool, you had a far easier time down there. Most of the time during a big-wave wipeout, I'd been told by Hamilton and others, the experience unfolded in a frightening but fairly predictable way. Once a rider had weathered the wave's impact, shaken like a rat in a dog's mouth for fifteen or twenty seconds, the energy eventually released him and he could make his way to the surface. The important phrase, however, was "most of the time." While some waves were forgiving, others seemed to have a distinct malicious streak. "It's the one-in-a-hundred wave you've got to watch out for," Lickle said. "The one that pins you on the bottom, stuffs you in a cave, and tells you, 'Son, here's a little lesson.'" Every big swell offered a chance to learn humility, to understand that what allowed a rider to go home with his spine in one piece was an easily blown cocktail of fate, skill, and attitude, with a twist of luck. Kalama had summed this up in the most straightforward way: "There is no guarantee that you'll be fine. You are completely at the mercy of the wave."

Lickle asked for my binoculars, and I passed them over. "There goes

Larry," he said, using Hamilton's nickname with affection. I looked out and saw Hamilton riding the nose of his standup board, twirling 360s and 720s. Dave and Ekolu had paddled so far offshore they looked like ants. Everyone else sat on their boards in the lineup, in much the same place they were the last time I looked. "Have you ever tried standup?" Lickle asked. "When it gets really gnarly out there, only Laird can do it." He handed the binoculars back. "He's not slowing down, you know," he said. "He's always gonna find something bigger and better."

"Bigger than Jaws?" I said. "What do you think it will be?"

Lickle was silent for a moment, and when he spoke his voice was lower. "We rode a wave one year down the coast," he said. "Solid eighty-footers. You always say, 'Is the hundred-footer ever gonna come?' Well, this thing was getting really close. And the bigger it was, the better it was. For years we never knew it was there."

"Where is it?"

Before he could answer, an exquisite seven-year-old girl walked up to us. It was Lickle's youngest daughter, Sky, a sweet-tempered kid with long chestnut hair, enormous eyes, and a splash of freckles across her nose. "Dad, are you going to be much longer?" she asked in a plaintive voice. Lickle ruffled her hair. "Just a minute, Sky-Pie. I'll meet you at the truck."

Sky wandered off and Lickle hesitated, as though he would prefer not to reveal the location. But then he told me. "When Spreckelsville is closing out, there's this *thing* going on outside of it," he said. "There's a wave out there that will literally hold the biggest swell of all. We call it Egypt because it looks like the great pyramids. Above a certain size, Jaws just gets thicker, like Teahupoo. So it will never hold the tallest wave in the world." He paused, and looked out at the ocean. "Egypt will."

Hamilton paddled his board all the way onto the sand, picked it up, and headed for the rusty outdoor shower to rinse everything off. A pack of kids trailed after him. In surf-speak these skinny, mop-haired urchins of the waves are known as grommets, or groms for short. Hamilton had a lot of time for them. Any older guy who bullied a grom in the surf wouldn't be doing it for long if he was around.

He stood washing the salt from his gear. On the sidelines, people circled with camera phones. "Maybe now the rest of us can get a few!" one surfer yelled as he walked by, laughing. "Hey," Hamilton said, smiling, "there's gotta be some reward for doing this for forty years." Two boys who looked like brothers hovered nearby, trying to summon the courage to speak. "That wave you surfed at Teahupoo is the sickest thing I've *ever* seen," one finally blurted out. "Yeah, sickest thing *ever*!" the other echoed. Hamilton looked up from rinsing his board, his eyes bloodshot from the sun. "You should go there," he said. "It's beautiful."

Beside him, a man in his thirties stepped out from beneath a large tree. "Laird," he said, "you probably don't remember my cousin, but he met you, he's a surfer and you were surfing . . ." His voice was high and tight and he spoke quickly, trying to rush into a connection.

Hamilton listened politely for a while but the man just kept talking, so he began to walk to his truck. The man followed, drawing his story to its punch line: "He hit you! My cousin! He scared the shit out of you!"

"Well, then I probably remember him," Hamilton said in a wry tone. He climbed into the driver's seat and lowered the window. Kalama was at the shower now, surrounded by groms. "Breakfast?" Hamilton yelled. Kalama looked over and nodded.

I headed out after Hamilton, driving down the Hana Highway and into Paia. If you turned onto the main street from the other direction, you passed a sign that said, "Welcome to Paia, Maui's Historical Plantation Town." Below that someone had affixed another sign: "Please Don't Feed the Hippies." The edict was delightfully impossible to obey, as everyone in Paia had a touch of hippie soul; it was only a matter of degree. No one cared about your résumé in Paia, or that you hadn't brushed your hair all the way through, or that your truck had seen better days. In the town's hub, a ramshackle grocery store called Mana Foods, yoga instructors shopped alongside heavily pierced drifters, and pot farmers mingled with supermodels, and Brazilian kitesurfers lined up at the deli counter behind Buddhist priests, and three-hundred-pound Samoan construction workers jostled in the aisles with movie stars, and everyone got along perfectly well. There was something about the town that brought the people in it down to earth, a pronounced antislickness. Nothing in Paia was sparkling new. Old posters peeled from wooden walls. Tin roofs looked like they'd been repeatedly pelted by hail. Built during the sugar boom of the late nineteenth century, the buildings had a faded, scuffed quality, even though they are painted in the most vivid colors. There were hot-pink buildings and turquoise buildings and lime green buildings. There was a vermilion-tinted building with canary trim. By contrast, Anthony's Coffee, Hamilton's favorite breakfast spot on the main street, appeared subdued with its pale green and white storefront.

Anthony's was run by Hamilton's friends Ed and Kerri Stewart, a couple from Seattle who knew a thing or two about coffee. It was an airy, white-walled café with ceiling fans, a mottled cement floor, and a blackboard menu written in rainbow colors. The place was usually packed, and this morning was no exception.

Hamilton walked past the line of people waiting to order and straight into the kitchen, where the cook, a hearty Hawaiian woman named Val Akana, met him with a bear hug. "Got some fresh ahi for you, bruddah," she said, as he continued on to the back patio, where he sat down at a rickety metal table. Within a minute Ed was there with Hamilton's regular drink, a quadruple long espresso. Ed was slightly built, with a brush of gray hair and a megawatt smile. He and Hamilton had a longstanding tradition of heckling each other. He greeted us and asked me what kind of coffee I wanted. "Do you want anything else?" he asked Hamilton.

"Yeah," Hamilton said. "I want to see you do a little tap dance for me."

"Got a pistol?" Ed said.

"I can get one," Hamilton said.

"You'd better." Ed smiled and began to walk away.

"That was real original, Ed," Hamilton said. "Think up something else and come back when you're ready."

Kalama arrived, followed by Lickle. Ed brought my Americano.

"Can I get an açai smoothie?" Hamilton asked.

"You know where the line is," Ed said with a smirk, before heading off to get the smoothie.

A waitress brought out platefuls of eggs and toast and lightly seared ahi with sides of salsa, avocados, and brown rice. As we began to eat, I asked Hamilton about an exchange I'd seen him having with a surfer in the lineup. It hadn't looked friendly.

"Do people get uptight in the water?" he said. "Totally. Somebody always has something to say. I think they get frustrated." The standup boards might have something to do with that, he conceded, but it wasn't an issue that concerned him. They could paste up all the go-away stickers they wanted—the oversize boards and paddles were here to stay. "It's the best training I've come across for big-wave riding," Hamilton said. "In a normal surfing situation, if you catch a long ride, that's twenty seconds. When you're standup surfing you're out there for two or three hours, working your legs, core, and foundation the entire time."

Kalama nodded: "It forces you to use your whole body, even the little muscles in your feet."

Hamilton continued, his voice rising. "And it's fun! Which means you're gonna do it a lot more. Unless you're just robot guy."

"Plus, you see things—it's like you're in a big aquarium," Lickle said, buttering his toast. "The other day I was out at Kanaha, between Lowers and the beach. I looked down and thought the bottom was moving; I looked again and realized, *'That's a fucking monster tiger shark.'*"

"They love to cruise on the bottom," Hamilton said. "They collect dead shit."

"He was in way shallower water than I ever would've imagined," Lickle said.

"You mean than you ever would've hoped," Kalama said.

"Well, we know they're out there," Hamilton said.

"Yeah, we do." Kalama fixed Lickle with a steely look.

Lickle laughed guiltily. "Hey, buddy . . . I said I'm sorry."

They were referring to an incident that had taken place a while back but had not been forgotten: "The time he trolled me," Kalama said drily. Lickle and Kalama had been tow surfing at Spreckelsville, Lickle driving and Kalama floating in the water on the end of the rope, waiting for a set. "I'm sitting on the Jet Ski looking off the back," Lickle recalled, "and I see this big-ass shark coming right at Kalama. It was far enough away . . . I just wanted to screw with him a little bit. So I said very calmly, 'Hey Dave, check out the size of that shark coming to visit you.'"

Kalama shook his head. "I see this dorsal fin—at first I thought it was a dolphin. But it just kept coming up out of the water and I realized, 'That ain't no dolphin. It's coming straight at me.' And this thing wasn't seven, eight, ten feet long. I want to call it at least fifteen."

"It was a big freakin' shark," Lickle agreed.

"So I yelled, 'Get me the fuck out of the water!'" Kalama said. "And he looks at me and laughs! He just starts laughing!"

"I go, 'No, have a better look at it!'" Lickle said, laughing. Hamilton was doubled over, laughing too.

"I'm thinking, *'You gotta be kidding me.'*" Kalama's voice was incredulous. "So I take it up a level. I yell louder. And he looks at me and laughs again. I'm like, *'What the* hell *is wrong with him?'*"

"That's grounds for disembowelment," Hamilton said.

"So finally I went to that bloodcurdling . . . *'I swear to God if I get out of this I am going to kill you!'*"

"And then I brought him up slowly," Lickle said, wiping tears from his eyes. "Just enough to get him to the surface."

"My ankles were still underwater," Kalama said. "But I'm moving. And now the shark's close—about fifteen feet away! And I'm like, '*What is wrong with this guy?*'"

"Well, the way I remember it," Lickle said, "I had a bit of humor going, and then he started playing the baby game so that made me—"

"Yeah, the *baby* game," Kalama said sarcastically.

"He was crybaby at the end of the line," Lickle continued. "I do remember just bringing him up to an ankle drag and going, 'Dude, check that big boy out!'"

"So did you get a good look at him?" I asked Kalama.

"Better than I would've liked! I just could *not* believe he did that. Here's my partner, I'm thinking, 'He's got my back in every situation.' And he's trolling me for a monster shark! And he's laughing about it!"

"And he's still laughing," Hamilton said, laughing.

"I really am sorry, Dave," Lickle said, trying not to laugh.

Looking around the table, I realized it would be hard to find a trio who had been through more together. They had staked their territory in an uncharted realm, a place where the ocean didn't necessarily allow people to be. The odd instance of partner trolling aside, they had saved one another's lives with chilling regularity. The reason why Hamilton, Kalama, and Lickle were all here now, still at the top of their games, with wives and kids and successful careers in a sport that did not dole those out easily, was because they did have one another's backs. As talented as each man was, the whole was more than the sum of the parts.

Far from instilling cockiness, their years of survival validated the attitude that had been there from the start: profound respect. Though none of the three had actually been born in Hawaii, they were native in their outlook, to the point of superstition. Whenever Jaws broke, they always carried a ti leaf along on the Jet Ski—a Polynesian tradition when going on a risky journey—for protection. "You take the leaf out," Hamilton explained, "and the leaf brings you home." For all the flash and technology that went along with tow surfing, they believed in timeless principles like karma, that a person gets back what he gives out, and they understood the hubris of humans trying to impose their will on the ocean.

Understatement was their way. A big-wave rider didn't exaggerate. He didn't hype his achievements or lose his bearings. The more amazing his feats were, the less he said about them. He could be confident, of course, but only losers ran around with what Hamilton called "puffy chest." As a matter of principle, the Hawaiians purposely diminished the size of a wave, measuring its height from the back rather than on its face. Thus, a twenty-foot wave was "ten-foot Hawaiian." "Usually the guys that do the talking are the guys that aren't doing the riding," Hamilton said. "Because if you've been in front of one of those waves [at Jaws], you don't flap like that. And if you do, you're asking for it."

In this belief system, to rush around after a cash prize for the title of First Man to Ride a Hundred-Foot Wave was to tempt fate. "As soon as Billabong put the golden carrot up, that was when the carnage started," Hamilton said. "That was the beginning of Skis on the rocks, guys getting hauled away. Everyone came out of the woodwork to get their shot at it."

"They didn't need to outsurf anybody," Kalama said. "They just needed to stand in the right place long enough for someone to take a picture."

" 'How big was my wave?' " Hamilton said, in a mocking whine. " 'Is my wave bigger than his wave? His was sixty-eight feet? Well, mine was sixty-eight and a half.' Biggest, longest, widest, tallest—what is it, a dog show?"

"Well, and number one," Lickle added, "if you're getting the prize for riding the biggest wave, you have to *make* the wave." He was referring to the Billabong XXL champion from a few years back whose winning ride had ended in a spectacular crash.

Hamilton leaned back in his chair and crossed his arms. The conversation was a reminder that another season was beginning, the fourth winter since anybody had needed to worry about measuring giant waves around here—there hadn't been any.

A shadow passed over his face. "I'll tell you what," he said. "All that contest stuff, that frenzy, that pursuit of the golden carrot—is what has caused Pe'ahi not to break."

"I believe that," Lickle said.

"It's done *something*," Hamilton said. "Because she's been very aloof since then."

"I think it had more to do with Brett riding the surf bike out there," Kalama joked. "Now that's disrespect."

"That was a contributing factor," Hamilton said, nodding. "Definitely."

"That had nothing to do with it," Lickle said.

I mentioned what Dave Levinson had told me, that the storm tracks were shifting, upending traditional weather patterns. Furious ocean conditions were out there, but they might not appear in the usual places or at the usual times.

"That seems right," Hamilton said, pointing out that the recent waves in Tahiti had come months out of season. "And we're getting a lot of strange swell directions," he added. "We're getting north. Or really west, almost southwest. Or superweird northeast. Not northwest, which is the usual window for the big winter swells."

The waitress came and cleared the table. The wind had picked up, sweeping a piece of palm leaf across the patio. Overhead, clouds hustled by as though late for an important appointment. Suddenly restless, Hamilton stood up to leave. "It's the calm before the storm," he said. "I just feel like there's only one thing that can happen when it's been flat. It's gonna get *real* big."

THE GLOBE BEGAN WITH SEA SO TO SPEAK;
AND WHO KNOWS IF IT WILL NOT END WITH IT?

Jules Verne, 20,000 Leagues Under the Sea

WAVE GOOD-BYE

LONDON, ENGLAND

No one understands the risks of an unruly sea better than Lloyd's of London, the British-based insurers of most of the global shipping fleet, a huge swath of the planet's most valuable real estate, and just about anything else you can think of. When a freighter disappears in the North Sea, Lloyd's pays. When storm waves surge into a low-lying city, Lloyd's pays. When an earthquake cracks the seafloor, sending a tsunami barreling toward a densely populated coastline, Lloyd's pays. There is nowhere a person could go to get a more exact reckoning of how dangerous and destructive giant waves can be than One Lime Street, Lloyd's headquarters in London's financial district.

I walked into the building, along with the morning crowds. One Lime Street is a glass and metal monolith nicknamed the "Inside Out" building because all of its workings—its cables and ducts and trusses and pulleys and vaults, all its stark iron guts—are visible. The lobby opens onto a twelve-story glass atrium crisscrossed with steel escalators that appear to be suspended in space. It was a long way from Lloyd's original headquarters back in 1688, a London coffeehouse where sailors and shipowners gathered to make impromptu insurance deals. When the sea proved to be every bit as lawless as sailors feared, the situations being insured against became legally defined as "Maritime Perils: the perils consequent on, or incidental to, the navigation of the sea, that is to say, perils of the sea, fire, war perils, pirates, rovers, thieves, captures, seizures, restraints, and detainments of princes and peoples, jettisons, barratry, and any other perils, either of the like kind or which may be designated by the policy."

Insuring ships was still at the heart of the operation, although over the centuries Lloyd's had branched out from its nautical roots, venturing so far afield that it had become known for insuring such valuables as Keith Richards's hands and Tina Turner's legs. Its policies covered the Golden Gate Bridge and, formerly, the World Trade Center in New York. Unique among its competitors, Lloyd's had proved willing to service unusual requests. One time, for instance, it insured a thirty-six-hour flight for ten elephants against "all risks of mortality." On another occasion, it agreed to cover a famous male flamenco dancer's pants against splitting midperformance. That's not to say that every client got the policy they were seeking: recently, Lloyd's declined to insure a two-headed albino rattlesnake because the last two-headed albino rattlesnake they'd had under coverage had died. The livestock underwriter's report reads with crisp finality: "An apparent disagreement between the respective heads had fatal consequences."

I had come to meet Neil Roberts, a senior executive who specialized in marine activity. Roberts, I hoped, would provide some perspective on the disappearing ships and what a stormier ocean climate might mean for Lloyd's business. How worried were they? In a story titled "Surf's Up: A Rising Tide of Natural Disasters," *The Economist* had reported in its 2007 preview issue that "the number of climate-related catastrophes tripled between the 1970s and the 1990s, and has continued to climb in the current decade." For insurance companies, this was an expensive fact. The aftermath of Hurricane Katrina, to cite just one example, had set them back more than $60 billion. "It is climate security which presents the biggest risk to insurers, and, for that matter, to us all," Lloyd's chairman, Lord Peter Levene, was quoted as saying. As I waited in the lobby I browsed through a booklet that outlined how climate volatility had prompted Lloyd's to deploy a supercomputer "the size of four tennis courts, three stories high, and housed in an earthquake-proof bunker," to create "realistic disaster scenarios" and assess the resulting claims exposure.

In short order Roberts appeared, a trim man with a broad, friendly face and just a hint of silver at his temples. Like everyone else in the building, he was immaculately dressed. Looking around at the dark suits, the smart ties, the below-the-knee skirts, it was clear that anyone looking to cut loose on Casual Fridays should search for another employer. Roberts and I shook hands and headed to a café in the middle of the lobby. Above us, floors of risk analysis, underwriting, and insurance trading thrummed with activity. We ordered coffees, and Roberts began to explain the kinds of harrowing situations that crossed his desk as a matter of course. His days were filled with troubles scrolling past like movie credits, bulletins streaming in with titles like "Climate Change Severe Threat to U.S. Coastlines," "Rapid Sea Level Rise," and "Piracy in the Gulf of Aden." I asked Roberts if Lloyd's considered freak waves to be a threat. "They have been on our committee's radar for a time now," Roberts said, sitting down at a table and opening a dossier of shipping statistics to show me. "They do exist. That's the one thing everyone agrees on." But in the grand scheme of maritime perils, he emphasized, giant waves—freakish or not—were among many concerns.

Since its earliest days Lloyd's had kept detailed records of ship losses in publications known as *Lloyd's List* and *Lloyd's Casualty Reports*. The original ledgers were archived in a nearby library; I'd spent previous days poring over their pages. From start to finish, they were testaments to the wisdom of staying on dry land. In a monthlong flurry in 1984, for instance, they recorded the loss of the *Marques*, "a British barque bound Bermuda for Halifax, knocked down by waves, flooded rapidly, sunk 70 miles north of Bermuda. Crew missing"; the *Perito Moreno*, an Argentine tanker that broke in two; the *Tesubu*, a Panamanian tanker carrying a cargo of molasses that began "taking water in No. 1 tank in heavy seas . . . Not sighted or heard of since"; the *Abu Al Khair*, a Kuwaiti cargo vessel that capsized in a gale and sank with all hands on board; the *Athena*, a nine-thousand-ton Greek ship "presumed to have sunk . . . after leaking in heavy seas"; the *Marine Electric*, a Panamanian ship "abandoned adrift with No. 1 hold flooded"; and the *Venus*, a Philippine motor ferry "sunk in heavy seas 110 miles southeast of Manila."

The ledgers were thick and dusty, oversize like showy coffee table books but with somber matte covers. Their entries were carefully inked in a courtly calligraphic hand on precisely lined paper, but their content was bare-knuckled and raw. I read through volumes of notations like "Continuous pounding by heavy seas broke her up" and "There was no sign of the other 13 men and because of the 50-foot waves further searching was impossible. It was presumed all had drowned." There were countless descriptions of tankers that had come apart at the seams: "The bow section drifted away in the storm tossed seas until it sank . . . some 40 miles away from the position of the stern."

The ships that met disaster most frequently were bulk carriers, a type of cargo ship developed in the 1950s to haul commodities like grain, coal, iron ore, cement, and timber. Bulkers were the Clydesdales of the sea, enormous steel workhorses. Three or four football fields in length, they sat low in the water and were flat across their decks. There was nothing supple or nimble about these ships, nothing that allowed them to do anything in big waves but lumber doggedly through them, take a heavy beating, and roll and twist and groan. In port they endured equal abuse: cargo was shot into their holds by high-speed machines, thousands of tons of iron ore pellets battering the boat's structure. During unloading, huge metal claws reached in and scraped the hull for every last ounce. Even the strongest steel would fatigue under these conditions—and corrode in salt water—and bulk carriers, especially those manufactured after 1980, were often made of something far weaker, a lighter high-tensile steel that cracked and corroded at an even faster clip.

Compounding these problems, bulk shipping's mammoth loads had an unfortunate tendency to shift in bumpy seas, causing fatal weight imbalances. While this usually meant that iron ore or a silo's worth of wheat was jouncing around below deck, a book that Lloyd's had published called *Modern Shipping Disasters* described the *Alexis*, a Greek cargo vessel carrying 2,500 sheep that "developed a list to port caused by the livestock becoming restless." The fidgeting sheep, it turned out, sank the ship. "A week later Agricultural Department employees began the task of burning or burying the rotting carcasses of 2,000 drowned sheep that had been washed up on western Cyprus beaches."

Bulkers have another Achilles' heel: their hatches, sizable deck openings that enable cargo to be shuttled in and out of the holds with maximum speed. (Time is money on the commercial high seas.) When waves pummel the deck, these hatches can be breached. If that happens, the hold beneath the hatch floods, causing the bulkheads between the various cargo compartments to blow out. From that point it is usually a matter of minutes before the ship goes to pieces.

The hatch problem was brutally illustrated on March 22, 1973, when two Norwegian bulk carriers, the *Norse Variant* and the *Anita*, disappeared at virtually the same time and in the same location. Both ships were carrying coal from Virginia to Europe when they ran into a storm off the coast of New Jersey, with fifty-foot waves and sixty-knot winds. The *Anita* and its crew of thirty-two disappeared so suddenly there was not even time for an emergency call, leading investigators to believe it was struck by a wave that arose from an unexpected direction, smashing open the hatches. No trace of the vessel was found. The *Norse Variant* sank with equal efficiency, although one man from among its thirty-one crew managed to survive. He was found two days later, barely alive and floating in the North Atlantic, 120 miles from where his ship went down. His account confirmed that the *Norse Variant*'s demise also had been caused by waves breaking a hatch cover, allowing water to flood in.

Often ship losses were blamed on metal fatigue and maintenance shortcomings—the vessel being too tired to withstand the waves' onslaught—and it was true that older ships tended to be more rust-buckety than younger ones, aging in a marine variation of dog years. Add to this the fact that bulkers were designed during a time when the biggest waves in the ocean had been written off as myth. It made no sense, there-fore, for anyone to build ships to meet such Leviathans. This mistake was made clear in 1980, when giant waves wiped out a ship that was far from old and decrepit, a state-of-the-art 186,000-ton bulk and crude oil carrier, the *Derbyshire*. It was the largest ship to be registered as missing by Lloyd's, and the largest British ship ever lost at sea.

Only four years old and well maintained, the *Derbyshire*, along with its crew of forty-four and its payload of iron ore, went down in the Pacific during Typhoon Orchid. No radio call went out, no distress signal was given. This wasn't a new story, perhaps, but unlike other lost ships this one wasn't flying a Liberian flag and manned by Laotian sailors. One would think that any 186,000-ton ship abruptly vanishing would be cause for thorough investigation, but in the past that was usually not what hap-pened. "The world has watched such catastrophes with icy detachment," British journalist Tom Mangold wrote. "There has been no widespread outrage, not even much demand for explanation. A handful of anony-mous Third World seamen drowning in some distant ocean cannot com-pete with emotive pictures of oil-covered seabirds off the Shetland Islands or polluted coastlines off Alaska."

In the *Derbyshire*'s case, things went differently. The crew's families formed a lobbying group and forced the British government to conduct not one but three inquiries. The results, at first, did not provide much comfort. "Forces of nature" and "poor seamanship" were blamed: the ship had been "overcome by 80-foot waves." The story might have ended there, had the families been satisfied by that explanation. They weren't. The *Derbyshire* was one of six sister ships built at approximately the same time, all with the same design, and by 1982 some of the siblings had exhib-ited alarming structural problems, including metal cracking so severe it emitted gunshot sounds. One of the ships had actually broken in half. Determined to uncover more details of what had gone so wrong so fast— and hopefully prevent it from happening again—the families and the International Transport Workers Federation raised money to hunt for the *Derbyshire*'s remains. Amazingly, they found them.

The wreckage was located 650 miles southeast of Japan, littered across the seafloor two and a half miles down. The stern had ripped off and lay more than a mile away from the rest of the pieces. Deep submersibles documented the scene with video and photographs, scanning a graveyard of metal that had been sheared and crumpled and torn. When the images were examined, surveyors discovered that, once again, a hatch cover had imploded near the front, causing water to flood the bow. But there was something else too. The type of metal fracturing on the hatches indicated that the ship had been hit by what engineers called "steep pressure impulses," a type of high-velocity impact that comes from plunging waves. In other words, it looked as though a wave had broken *on top* of the *Derbyshire*. And while an army of eighty-footers would assault any ship, it would have required a greater beast than that to deliver a knockout blow from above. Some investigators believed that only a freak wave could have done it, with its abnormally steep, breaking peak and its profound trough for the ship to fall into. Of course, there was no way to prove this. "This is still a casualty without an eyewitness," the British attorney general Lord Williams declared, summing up the obvious: "Those who were on board the *Derbyshire* were the only persons in a position to know what happened, and all of them have perished."

After a decade of effort, the families of the lost crew had succeeded in drawing attention to bulk carriers' sketchier features, such as hatch covers that were too flimsy to meet extreme waves, and other problems long overdue for fixing. Regulations were tightened, stricter safety procedures advised. There was a chorus among naval architects that, given the wide gap between what the models had pronounced about maximum wave size versus what Nature had to say on the subject, ship design needed more than a little tweaking: it needed a total rethink. "It is true that the loss of the *Derbyshire* prompted big changes," Roberts said.

Not big enough, unfortunately. According to the UN's International Maritime Organization (IMO), from 1990 to mid-1997 a total of ninety-nine bulk carriers were lost. Then, in a dire encore, twenty-seven vessels along with 654 people were lost during a four-month period in the winter of 1997–98. Oil tankers slipped from the radar, leaving only black slicks to show they'd ever existed; rescuers responding to emergency calls arrived at the coordinates and found, instead of the vessel, mangled bits of debris. "In some cases ships had simply broken apart like a snapped pencil," an IMO report read.

During the twenty-first century, ships have continued to submit to the waves at startling rates. Paging through loss reports from the International Union of Marine Insurance (IUMI), Roberts pointed out that statistics might not tell the whole story, given the industry trend toward building ships so gargantuan that special ports must be constructed to accommodate them. (One vanished ship now is kind of like two in the past.) Stormier seas seem to be taking their toll: in recent years losses caused by weather have risen by more than 10 percent.

Cargo ships weren't the only ones running into trouble in the swells. Roberts also worried about new trends in the cruise ship industry, notably the ballooning size of the vessels and itineraries to increasingly far-flung ports of call. Engines could fail in remote areas where rescue was impractical or even impossible. "You've got five thousand people on some of these cruise ships," he said. "It's a high concentration of risk." In recent years there had been numerous incidents—from pirate attacks to ship-crippling fires to hellacious storms—where cruises had been anything but carefree vacations.

In 1995 the Cunard liner *Queen Elizabeth 2*, a thousand feet of plush carpets, grand pianos, and soft lighting, was buffaloed by a pair of ninety-five-foot waves that jumped out of sixty-foot Atlantic seas churned up by

Hurricane Luis. The captain, Ronald Warwick, was able to determine the waves' height when they "loomed out of the darkness from 220°" because their crests were level with the bridge. (Ocean buoys moored nearby recorded even higher waves at the time.) "It looked as though the ship was heading straight for the white cliffs of Dover," Warwick said. The waves broke over the bow with explosive force and the ship fell into the trough between them, smashing many of its windows, part of its foredeck, and, one would imagine, all of its crystal stemware. Amazingly, the *QE2* survived the encounter.

Lindblad Expeditions' *Endeavour* had its navigation and communications equipment and its windows blown out by a hundred-foot wave near Antarctica but still managed to limp to safety; Holland America's flagship *Rotterdam* lost power in four engines in the middle of Hurricane Karl, leaving the sixty-thousand-ton vessel floundering in fifty-foot seas for six hours until backup generators kicked in. In 2007 the 2,500-passenger *Norwegian Dawn*, en route from the Bahamas to New York City in heavy seas, was struck by a seventy-foot wave that smashed windows, flooded cabins, ripped Jacuzzis off the decks, and hurled passengers from their beds. "The sea had actually calmed down when the wave seemed to come out of thin air at daybreak," a spokeswoman said. "Our captain, who has twenty years of experience, said he'd never seen anything like it." Despite offers of free drinks and discount vouchers for future cruises, not every passenger was mollified. A group of them filed a lawsuit against the cruise line, claiming that the freak wave should have been "reasonably expected."

We finished our coffees and Roberts offered to give me a tour of Lloyd's underwriting section, famously known as The Room. It was one level up the crazy escalator and, in fact, wasn't a room at all. It was a long open space, a hive of brokers, insurers, and clients, buying and selling and totting up risk. Like Nasdaq or the New York Stock Exchange, Lloyd's is a market; this was its trading floor. Standing quietly in the center of the action was the Lutine Bell. Majestic in its twenty-foot-high circular pavilion made of dark, heavily polished wood, the bell had belonged to a frigate that sank in 1799 with a cargo of gold and silver bullion. Over the years it has served as a combination centerpiece and mascot in every Lloyd's headquarters. Traditionally it was sounded whenever a ship went missing. Now it was rung only ceremonially, tolling once for bad news, twice for good. On September 11, 2001, Roberts told me, the building heard its single, mournful peal.

Roberts, a nautical history buff, led me to a glass display case near the bell containing, among other artifacts, the original logbook from the Battle of Trafalgar. It was an astonishing thing to see, set among hundreds of blinking computer screens on the trading floor, relics from the oceans of the past transported centuries into the future, reminders that although humans and their ships might come and go, the seas will always remain.

Next to the case, two large ledgers lay open on a table. They were identical to those I'd seen in the library, except that the one in front of me bore the current date. Its companion ledger was labeled with the same day and month—but from one hundred years ago. I was staring at these books, both of which were filled, like all the others, with ocean mishaps, when a startlingly tiny man walked up, leaned over the current ledger, and began to carefully write in the familiar calligraphy:

Northsea
Cambodian ship
Sank following a fire on board in Lat 04 44N
N long 0234 W
22 crew rescued
4 dead, 3 missing

As he inscribed the loss his face was serious, and when he was finished, he nodded at what he had written and walked away.

Shipping will continue to be treacherous, Roberts said as we made our way back to the lobby. There were always fresh worries, the latest ones including a global crew shortage. This lack of expertise was especially troubling given the next-generation ships, floating colossi with complex computer navigation systems to master, not always a snap when the manual's written in German and you speak only Tagalog. "The number of adequately experienced mariners will be spread even more thinly," Roberts noted.

Along with the classic marine perils of yore, now Lloyd's also had to weigh modern risks of terrorism, pandemics, cyberattacks, and climate volatility. Meanwhile, they expected not only snarlier oceans and elevated sea levels, but more hurricanes, windstorms, storm surges, floods, earthquakes, wildfires, droughts—all affecting more people and more property. I could see what Roberts was trying to explain. Yes, ships had it rough out there, and sure, the losses were astonishing, but these were dwarfed by Lloyd's nightmare scenario: a disaster impacting the eastern seaboard of the United States, where over eight trillion dollars' worth of coastal property, 111 million people, and half the U.S. gross domestic product would be exposed. A larger-scale repeat of Katrina or a steep sea level rise, or even—as outlandish as it sounds—a tsunami would make things like freak waves and missing cargo ships seem pretty unimportant.

I thanked Roberts and left the Inside Out building, exiting into a brilliant, windswept afternoon. Behind me the risk business continued its work and its bets—for or against—the continued well-being of celebrity body parts, valuable ships, and vulnerable cities. I walked back to my hotel, realizing that there was someone else I needed to see in London. As I was leaving Lloyd's, Roberts had asked: "Have you been to Benfield Hazard Research?" When I said no, he raised an eyebrow. "Bill McGuire," he said. "He's someone to talk to about waves. I heard him speak at a conference. Now there's a man who knows how to worry all of us."

The University College London campus sprawls through the center of the city, blocks and compounds of classic, if slightly sooty, Georgian buildings. "College of this, college of that," my taxi driver said dismissively, pulling up to the Lewis Building on Gower Street. It was an academic-looking place, deceptively peaceful given the chaos that was being studied behind its staid limestone facade. After trailing through a maze of corridors, I found the entrance for the Benfield Hazard Research Center. The center (now called Aon Benfield UCL Hazard Research) specializes in analyzing and forecasting geological upheavals like tsunamis or earthquakes, and meteorological hazards such as hurricanes and floods. A red sign on the wall said: "Reception Area: Please Do Not Enter."

As Roberts suggested, I had come to meet Bill McGuire, the center's director, a prominent volcanologist, geophysical hazards expert, and media personality whose predictions of biblical-scale natural disasters had earned him the nicknames the Prophet of Doom and Disasterman. In his lectures, scientific papers, and radio and television appearances, McGuire routinely set forth a buffet of unappetizing future scenarios that all corresponded to the same theme: *watch out.* "So far we have prospered," he had written in his book *Apocalypse*, "but the greatest battles with Nature are yet to be fought, and the final outcome remains in the balance." The author of several popular books, he had a way of outlining the most horrific disasters in jaunty, accessible prose. "The big problem with predicting the end of the world," he'd mused, "is that, if proved right, there can be no basking in glory."

McGuire's stock in trade was what he referred to as "Gee-Gees," short for global geophysical events. To qualify for this designation, a natural disaster had to have a widespread and fearsome impact. It had to rattle societies and upend economies and claim enormous numbers of victims. In his lineup of double-barreled catastrophes McGuire had a lot to say about waves, unimaginably large waves.

The receptionist rang McGuire's extension and he came downstairs. I was disappointed because I had hoped to see his office, imagining it as a kind of preapocalyptic mission control with satellite feeds, cyclone statistics, and real-time seismographic readings flashing across wall-size digital maps. As for McGuire himself, I guess I had envisioned him as a brooding character. In reality, the man who showed up in the reception area was buoyant, with light hazel eyes and a round, bemused face. He wore faded jeans, a striped button-down shirt, and small glasses with pale frames. As we walked to one of the college's canteens for lunch, I found it hard to reconcile McGuire's cheery presence with the subjects of his research, things that might have come straight from the Book of Revelation: volcanic explosions, asteroid strikes, million-fatality earthquakes, and thousand-foot megatsunamis thundering across entire ocean basins. It would be comforting to write him off, but McGuire's credentials made that impossible. Crackpots weren't often invited to speak at Lloyd's of London or run respected research groups or publish scientific papers in *Nature*.

We turned into a Gothic building that could have been a library or a lecture hall but in fact was a pub. "I think I'll have a pint," McGuire said, settling into a chair. I ordered one too, and when the waitress brought them over, I made a toast. "Here's to being here," I said, "and not underwater or something."

"Not yet, anyway," McGuire said flatly, taking a drink.

Early in his career McGuire, a postgrad geology student, had been hoping to work on the Mars Viking Lander program, but a posting in Sicily studying Mount Etna sold him on volcanology instead: "It was three years of eating pasta and drinking lots of red wine." After that he was deployed to Montserrat, a British territory in the Caribbean West Indies where a volcano called Soufrière Hills sprang violently to life in 1995 after three centuries of dormancy. British volcanologists were in short supply, and McGuire found himself the senior scientist at Soufrière during its most dramatic eruptions, one of which happened on his second day there. "If it had been any bigger, we'd have all been dead," he said. "We didn't know what was going on. We were just incredibly lucky."

Soufrière belched out lava fountains and yellow-gray clouds of sulfur that choked the air. It sent pyroclastic dervishes of ash, gas, and burning rocks hurtling down the volcano at one hundred miles per hour. Eruptions buried the capital city, Plymouth, under forty feet of mud and turned the day as black as night. Instead of rainfall there was "ashfall." The island's vegetation died, as did nineteen people who didn't get out of the volcano's way fast enough. When Soufrière started its engines, life as Montserrat's inhabitants knew it ended. The fabric of the little society, its tourist trade, its economy, its way of life, all were smothered under a red-hot blanket. For McGuire, it was a graphic lesson in just how destructive nature could be.

Unless you've been through it, this level of mayhem is an abstraction, a nightmare that lodges itself in a dark corner of your mind but never really gets taken seriously. Despite detailed accounts of outlandish catastrophes from recent centuries, there's a collective amnesia about them. As the global population clusters along the coastlines, the cautionary tales of the past—the oceans that were three hundred feet higher, the lost civilizations, the sunken islands, the redrawn maps—are long forgotten. "We have yet to experience the blind terror of being faced with a sea wave higher than a cathedral," McGuire wrote. "Consequently these threats have no meaning for us on a daily basis." Yet it was only 124 years ago that Krakatoa, a volcanic island perched between Java and Sumatra, blew its top and partially collapsed, creating a 140-foot tsunami that struck at eighty miles per hour, wiped out 165 villages, killed 36,000 people, and barely slowed down. (The first report of the disaster was a Morse code message sent by a Lloyd's agent stationed on Java.)

So few people have ever seen or documented the epicenter of a huge tsunami—or lived to tell about it—that it's hard to envision how it happens or what it is. Contrary to the way it's usually described, a tsunami consists not of a single, malefic wave but of a series of them. Generated by earthquakes and landslides that thrust the water vertically upward or downward, it's an immense spasm of energy that reaches all the way to the ocean floor, and when it hits land it unleashes everything it's got. "The surface of the sea had a terrible, writhing, coiling awfulness to it," one Krakatoa witness described, adding, "I am convinced the Day of Judgment has come."

You don't have to go back very far in history to find even more calamitous incidents. McGuire points to an event, pegged to around A.D. 365 and known by the disquieting name "The Early Byzantine Tectonic Paroxysm," during which earthquakes thrust country-size hunks of land thirty feet in the air, generating a tsunami that steamrolled much of the eastern Mediterranean coastline. In his book *Apocalypse*, he writes: "There is absolutely no reason why such a readjustment of the complex Mediterranean geology could not occur again."

In 2000 McGuire noted in a newspaper column that things had been eerily quiet on the tsunami front, predicting that this would change, particularly in Indonesia. Four years later he was proved horribly right: a magnitude 9.1 earthquake tore across 740 miles of the Indian Ocean near Sumatra, punching a piece of the seafloor sixty-six feet upward and tearing open a thirty-three-foot rift. The quake—estimated to have contained the energy of 23,000 Hiroshima-style atomic bombs—shook for a full ten minutes, setting in motion a hundred-foot tsunami that obliterated the city of Banda Aceh on Sumatra's northwestern tip (the closest city to the quake's epicenter) and then continued on at a smaller, but still devastating, forty feet to other parts of Indonesia, India, and Africa. The three main waves killed 240,000 people, left two million homeless in more than a dozen countries, and destroyed everything in their path. "I have been in war, and I have been through other relief operations," then–Secretary of State Colin Powell said, visibly shaken. "But I have never seen anything like this." He wasn't alone. The waves' power caught the world by shocking surprise.

"Tsunamis are no ordinary waves," McGuire said with understatement. "They are walls of water that just keep coming in. If it's one hundred feet high, it's going to be one hundred feet high for five minutes." As rare and unlikely as they sound, tsunamis—a Japanese word that loosely translates to "harbor waves," because they become visible only when they near land—are as inevitable as hurricanes or floods. The Pacific Ocean alone produced nearly a thousand in the past century. Depending on the geological event that caused it, a tsunami can measure anywhere from an inch to more than a mile high when it stampedes ashore. Japan has been walloped twenty-five times in the past four hundred years, with deaths in the hundreds of thousands. Tsunamis every bit as powerful as the one in 2004 have inundated America's west coast sixteen times over the past ten thousand years, most recently in 1700. Smaller waves—still lethal and destructive—appear in the Pacific Northwest, Alaska, and Hawaii far more often. Which is not surprising when you consider that the Pacific basin, a patchwork quilt of tectonic plates grinding against one another, is an earthquake factory.

When a quake or a volcanic eruption does more than jostle the seafloor, when its motion causes an underwater landslide or shakes loose a chunk of coastline or glacier, the resulting waves can measure not in the hundreds but in the thousands of feet. Though we don't often think of them this way, the oceans are filled with mountain ranges, trillions of tons of underwater rock and lava that shift around as time goes on. Volcanic islands—piles of loosely aggregated material heaped up by successive eruptions—are especially precarious. The steeper they grow above the water, the faster the ocean erodes them from beneath, and eventually they all topple over. "They've detected seventy huge collapses in Hawaii," McGuire said. "There are tsunami deposits on the islands that show that the waves have been at least 170 meters [550 feet] high."

Fortunately this cycle takes place over millions of years, and only the most paranoid among us would actively fear such a disaster. But if you wanted to bet on which volcanic island would next crumple into the sea, McGuire has a contender for you: La Palma, in the Canary Islands. It's the steepest island in the world, and one of the most volcanically active—not a comforting combination—and lately its Cumbre Vieja volcano has been demonstrating some fairly alarming behavior. It has erupted seven times, most recently in 1971, and is poised for an eighth.

What's more, previous eruptions have already caused a west-facing section of land to drop fifteen feet lower than the rest of the island. The two sections are divided by a fissure, a weak spot where the rock has split. When McGuire's colleague, British geologist Simon Day, came to take a closer look, he found even more bad news: the twenty-thousand-foot volcano itself was bisected by a fault line, and the inside of the crater was loaded with water. "Volcanoes act like giant sponges," Day explained, "and that weight creates an unstable situation." When magma is present its heat turns that water into steam, which can then blow apart sections of rock. If Cumbre Vieja's next eruption triggers a slide, this house-of-cards island could shed its entire western flank—about thirteen miles long and ten miles wide, a mile thick, and weighing approximately 500 billion tons.

When Day analyzed his research, he was stunned: the model showed that when this land dropped into the ocean, the resulting splash wave would be three thousand feet high and generate a tsunami that would hit the Canaries, the northwest coast of Africa, southern Europe, the U.K., the Caribbean, North and South America. By the time it arrived at the U.S. East Coast nine hours later, the wave would still top one hundred feet.

While McGuire and I talked the lunch crowd poured in, though it was clear no one had come for the food. I glanced at a tray of brick-heavy meat pies and hot dogs curling under a heat lamp and ordered another Guinness. McGuire did the same. "The earth fights back is what I tend to say," he said. "Climate change is *the* greatest Gee-Gee of all time. If we don't tackle that, we're not going to be in any position to tackle the rest of it." While he acknowledged the impact of increased storminess—"Wave heights around the U.K. have increased by about a third in the last few decades"—McGuire emphasized that climate change has additional wave-generating effects that few people are aware of. "If you start to see meter-scale [3.3-foot] rises in sea level, then that load starts to bend the [earth's] crust, and that would promote magma reaching the surface. That will give you a massive increase in volcanic activity. It'll activate faults to create earthquakes, submarine landslides, tsunamis, the whole lot." As bizarre as this sounds, history—and other scientists—back up his theory. "Many potentially hazardous geological systems are sensitive to changes in currents, sea level, and atmospheric pressure," NASA geophysicist Dr. Jeanne Sauber said in a *New Scientist* article. "It's unavoidable that glacial retreat will induce tectonic activity."

Sitting calmly in his chair, sipping at his pint, McGuire spoke about Florida being underwater, an asteroid splashing down in the ocean (now *there* would be a wave), and an earthquake wiping out Tokyo. The Caribbean (particularly Puerto Rico) and the U.S. Pacific Northwest were probably overdue for tsunami-inducing quakes. "It's complicated," he said, "but generally speaking, if you warm the earth up very, very rapidly, and you're pumping more energy into the weather machine, you're going to see more dynamic events of all types. It's exciting, actually." He laughed nervously, and then corrected himself. "Probably not the right word, *exciting*."

Earlier McGuire had mentioned his four-year-old son, Fraser. I wondered how he balanced his fears of runaway climate change and wrath-of-God natural disasters with his hopes for Fraser's future. "Well, I think his life will be much harder than mine has been," McGuire said matter-of-factly. An introspective look came over his face. "The world is undoubtedly going to be a much more difficult place for him when he grows up." He paused. "People ask me how I sleep at night," he said. "And I tell them, 'Like anybody else.' I can't lie there thinking, 'Oh my God, there might be a supereruption tonight.' It's not human nature. But in the daytime I will consider things like that."

Despite his dire prognostications, McGuire considered himself an "optimistic pessimist." The idea of being a "prophet of doom," frankly, bummed him out. As he saw it, what he was doing was simply pointing out the facts: it may not happen tomorrow or even ten thousand years from now, but this stuff *does* happen. A mile-high asteroid wave erupting in the Pacific—it wasn't something he'd made up. Those sunken, blasted-out volcanic craters that dotted the oceans? At one point they had looked a lot like Hawaii.

I had a hard enough time envisioning even a hundred-foot wave, I told McGuire. How on earth do you conjure up a wave ten or twenty times that size in your mind's eye? Landslide-induced giant waves, he explained, trying to give me the picture, didn't start off looking like waves. They were more like aquatic mushroom clouds. "The water just sort of bubbles up into a wave," he said with a chuckle. "It isn't something you'd forget." A mile-high wave seemed completely otherworldly, except that it wasn't. In fact, there was a place in North America that specialized in generating them. And there were three people alive who, in a manner of speaking, had surfed one.

YOU'RE PLAYING WITH THE EDGE HERE.

Big-wave surfer Jeff Clark

MAVERICKS

HALF MOON BAY, CALIFORNIA

Hamilton's storm prediction proved right. As December began, the weather radar screens pulsated with the mightiest magenta blob anyone had seen in years, as a mammoth disturbance snaked its way across the North Pacific. A cold low-pressure system had joined forces with a warm low-pressure system, the extra heat and moisture whipping the two storms into one howling monster. "The Northern Hemisphere is going absolutely ballistic right now," Surfline reported. (Similar furies were also under way in the North Atlantic, with fifty-foot waves lashing the coasts of Ireland, England, France, and Spain.) Sometimes the magenta blobs started off like they meant business only to fade in the end, but for giant wave potential this storm looked solid. There was only one problem: it might be a little *too* solid. Conditions might be too haywire for the waves to be rideable. This was a full-on cyclone and it was traveling from an unusual direction, west-southwest. Typically the North Pacific storms rumbled down from the Bering Sea at a northwesterly angle. This one had dipped farther south and looked like it would largely sidestep Hawaii, barreling directly toward northern California and Oregon. A huge swell was coming, that much was clear, but the wind conditions would determine whether the waves were mad, messy heaps of water or the glassy clean skyscrapers that big-wave surfers dream of.

Forecaster Sean Collins monitored satellite and buoy readings, wind speeds, wave spectra, and model predictions, considered the numbers, and consulted LOLA—Surfline's custom computer model that filters sea state data through a surfing prism. He arrived at his verdict late on December 2: the swell's most desirable waves would be found at a northern California break called Ghost Tree, on the morning of December 4. E-mails went out, plane tickets were booked, Jet Skis were corralled, and from Hawaii to Brazil to South Africa, the riders snapped into action.

I had been traveling when I heard about the swell, headed to Los Angeles from the East Coast. From L.A. I caught the last flight to San Francisco on December 3, planning to drive the 125 miles south to Ghost Tree the next morning before dawn. The break, improbably located about a three-iron shot off the eighteenth hole at Pebble Beach, the famous golf course near Carmel, was named after a dead cypress husk on nearby Pescadero Point. Among big-wave connoisseurs, Ghost Tree wasn't especially beloved. It didn't break that often, and when it did it lunged open in a maniac sneer, spitting foam and tangled rafts of kelp. A minefield of rocks fringed its base, leaving surfers no margin for error. Boils, seething disturbances in the water that indicated a shallowly hidden obstacle beneath, burbled up all over the place. Ghost Tree was a monster truck of a wave, huge and showy and growly but not especially comfortable to ride. It had one advantage for this storm, however: the deepwater canyon that created the wave was ideally angled to capture a west swell.

I called Hamilton to see what he was planning, but his cell phone had clicked straight to voice mail: *"Due to the submersion of my phone, I no longer have your number,"* his message said. Cell phones didn't last very long around Hamilton. They got crushed under the wheels of trucks or forgotten at the hardware store or lost in the pineapple fields or, it appeared, dropped into the ocean. I left a message but wasn't sure when I would hear from him: Gabby Reece, now fully nine months pregnant, was due to have their baby any day. It seemed unlikely that he would consider leaving Hawaii right now.

Even if he hadn't been awaiting the arrival of his third child, Hamilton was not typically a participant in the seat-of-the-pants global big-wave hunt that kicked into *Amazing Race* mode whenever a promising magenta blob showed up. Tahiti was an exception; he had a special connection to the wave, a support system in place, and a couple of jet Skis stashed on the island. "Chasing the rabbit is a tricky game," he'd told me one time, explaining his philosophy. "You really don't want to get into that at a certain point. You're gonna be looking over there when it's over here and running around half-cocked and unprepared when it does happen." His usual strategy was to stay put, in the location that was most likely to pay off over time: "There's a reason why surfing began in Hawaii, you know."

Still, there were waves on the West Coast that intrigued him, notably the Cortes Bank, an offshore break located one hundred miles out in the Pacific, due west from San Diego. The waves at Cortes were created by an underwater mountain range that rose five thousand feet from the ocean floor to within six feet of the surface, in spots. Many people believed Cortes was the likeliest bet to produce a rideable hundred-foot (or even hundred-fifty-foot) face, although given how exposed the place was, its optimal weather conditions required the wind, ocean, sun, moon, and stars to align. Even then, you weren't just going surfing at Cortes, you were going on an expedition. Hamilton had also expressed interest in Mavericks, a foreboding wave just thirty miles south of San Francisco. In any case, I doubted that Ghost Tree would be much of a draw. One time I asked him what he thought of the place. "Big waves are all beautiful in their own way," he said. "But I'll tell you one thing, it ain't Pe'ahi."

After I landed in San Francisco, I called Sean Collins. He was already down in Carmel. "Ghost Tree should be huge," he said. "It's a really, really big swell. I think Mavericks is going to be big too, but it might have south wind problems. Considering the weather, Ghost Tree will be cleaner." He gave me directions to the best vantage point, which happened to be on private property and was thus a closely guarded secret.

While we were speaking, another call beeped on my phone. I finished talking to Collins, and then listened to my voice mail. The message was from Mike Prickett, a filmmaker who was flying in from Oahu for the swell, along with a contingent of tow surfers and photographers. They were going to Mavericks, Prickett said, and they had arranged for a boat. If I wanted to watch the action from a wave-side seat, there was plenty of room. It was an easy decision: the promise of getting out on the water trumped Ghost Tree's better wind forecast. I called Prickett back and accepted his invitation.

"Thirty-two feet at twenty seconds!"

Garrett McNamara, yelling like a stockbroker, twisted around in the passenger seat of my rental truck and thrust his iPhone at Kealii Mamala, his tow partner, sitting in the backseat. Mamala, a striking Hawaiian with a nimbus of curly brown hair, looked at the buoy reading on the screen, and smiled. "Oh, *yeah*," he said. A thirty-two-foot swell with a period that long meant sixty- and seventy-foot waves and beyond; it was a deep-reaching blast of power, an emissary sent to deliver the ocean's most humbling message: today the riders would be playing a role that Hamilton liked to describe as "Ant Man."

We were ten miles north of Half Moon Bay, a quiet fishing town that was the launch point for Mavericks. The skies enclosed us in a shroud of gray drizzle, turning everything dark despite the fact that it was seven-thirty in the morning. Fog slunk along the edges of the road. Though I couldn't see it, I knew the ocean was close by and that somewhere beyond the gloomy curtain, the swell was marching toward us. I flicked the windshield wipers on and wished that the word *ominous* didn't keep popping into my head. "I think it's clearing up," McNamara said hopefully, pointing to a slightly less glowering sliver of sky.

At five o'clock that morning I had gone to the San Francisco airport to meet the night flight from Hawaii. I caught up with Prickett in the baggage claim, where, like all the photographers, he was wrangling heavy cases of gear and still half-asleep. His light brown hair stuck up at the back where it had been plastered against the airplane seat. If you didn't know he was one of the top ocean cinematographers, you'd take Prickett for one of the surfers. He had the same disheveled cool, a hint of a hell-raising look in his eyes, and a movie star smile. Standing with him was Tony Harrington, an Australian photographer whom I had met in Tahiti. Harro, as he was known, was another celebrated name behind the lens. He specialized in dropping in on the wildest weather—in the mountains as well as the ocean—and his more extreme exploits had been made into a TV series called *Storm Hunter*. He was a tallish guy with a rugby player's heft, blond-haired and round-cheeked and friendly as a golden retriever, but when the situation turned intense, so did Harro. I saw Garrett McNamara gathering his boards and went over to say hello. He was wearing a green hoodie and the same intense expression I remembered from Teahupoo. Due to some mix-up, he and his tow partner, Mamala, needed a lift to the wave, so I'd volunteered. It seemed like a small favor. They had spent the previous day paddle-surfing twenty-five-foot faces at Waimea Bay, traveled through the night to get here, and now, on maybe two hours of sleep, they were about to launch themselves into the heart of the magenta blob.

As we drove I asked why they'd chosen Mavericks, given Collins's recommendation of Ghost Tree. In Tahiti Collins had called the swell down to the hour, so I was somewhat nervous about bucking his advice. On the other hand, I knew that McNamara and Mamala had their own antennas up. They hadn't flown all night to come to Mavericks because they thought the waves might be better somewhere else.

"Ghost Tree is a shitty wave," McNamara said. "It's this big rolling thing and then you end up on the rocks if you screw up. And it's not like Mavericks' rocks, where you can escape through a hole. You're pounding right on the cliff."

"That's a wave you don't want to fuck around with, man," Mamala agreed. "It's basically all chop down the face: *cha-cha-cha-cha-cha.*" He made a noise like a machine gun, like teeth rattling. Mavericks, he added, was his favorite wave, "because of the dangers. When you go to Mavericks you're like, 'My God—sharks, cold, this, that.' And my first time there I got *worked*. But now I love it."

If, as the surfers claimed, every big wave has a distinct personality, Mavericks was an assassin. While other waves glimmer in the tropical sun, Mavericks seethes above a black chasm. Perched just north of Monterey Bay's abyssal canyons, its surface is as impenetrable as one-way glass. The Aleutian swells thunder three thousand miles across the North Pacific, barging past the continental shelf until their progress is rudely halted by a thick rock ledge that juts offshore about a mile from Pillar Point, near Half Moon Bay's harbor. When it hits this shallower depth, the wave energy rears up, shrieking and screaming, forming the clawed hand that is Mavericks. Around here the water temperatures hover in the low fifties, making everything harder—literally. Cold water has a higher viscosity. It is thicker, like liquid pavement, compounding the brutality of a fall. Frigid temperatures also make it tougher for surfers to relax, to paddle, to hold their breath underwater, to keep their senses from numbing over in general. The year-round uniform at Mavericks is head-to-toe neoprene, including hoods, boots, and gloves, which restrict the riders' movements and make it harder to feel the wave's gyrations. "I think of my feet the way other people think of their hands," Hamilton had told me, explaining how important that was for control. But that kind of sensitivity wasn't possible when there were five millimeters of rubber between a rider and his board.

If all this weren't daunting enough, Mavericks was located at the southern end of a region known as the Red Triangle because more attacks by great white sharks had occurred there than anywhere else on earth. Surfers had been bumped, bitten, and killed in nearby waters; sitting or paddling on their boards, clad in their black wetsuits, they resembled nothing so much as seals, the white shark's main prey. On at least two occasions at Mavericks, surfers had been catapulted into the air on their surfboards when sharks charged them from below. Down by Ghost Tree a rider had disappeared, never to be seen again. Later his board washed up onshore; it was punctured with bite marks that matched the jaws of a twenty-foot shark. But while great whites hadn't taken the life of any surfer at Mavericks, the wave itself had.

On December 23, 1994, one of Hawaii's best-known big-wave riders, Mark Foo, had flown over to Mavericks for a swell, made what appeared to be a fairly standard fall on a thirty-foot face, and failed to surface—for an hour. Other riders saw the tumble, during which Foo's board snapped into three pieces, but in the frenzy of the day nobody registered his absence until it was too late. The surfers were paddling into the waves, not towing, so Foo had no partner focused on his safety. When he didn't reappear in the lineup, everyone presumed he had gone back to shore to get another board; it was only when his body was found floating near the harbor that the truth became clear. Afterward people speculated that Foo had hit his head on the bottom and blacked out, or that his leash snagged in the rocks, trapping him underwater. But it was also possible that he drowned in a merciless set-long hold-down, the wave simply refusing to release him.

His death was a tragic validation of Brett Lickle's theory that every so often a wave came along that was meaner than others, and that fate was part of the equation: Foo was an ace who had surfed many waves far larger than the one that killed him. But he was from another quadrant of the Pacific; uneasy, perhaps, in his full-body wetsuit, jet-lagged from his flight, and facing Mavericks' wolfish pit for the first—and last—time. Other surfers were unnerved by a saying that Foo had recited so often it became one of his hallmarks: "If you want to ride the ultimate wave, you have to be willing to pay the ultimate price." Horribly, he did.

Along with the wave itself, Mavericks' surrounding waters were tricky and shifty and given to evil behavior. During storms in this neighborhood, the ocean's energy could flare. McNamara and Mamala recounted the story of their friend Shawn Alladio, a water-safety expert who had encountered a series of surreal waves outside Mavericks on November 21, 2001, a day that became known as "Hundred-Foot Wednesday."

The day had started out imposingly enough, but it intensified dramatically as multiple storms moved in. Patrolling on Jet Skis, Alladio and her colleague Jonathan Cahill spent the morning gathering lost boards, helping stranded surfers, and performing rescues. By early afternoon the conditions had become too nuts for anyone to be out, and even the tow surfers went back to shore. About four hundred yards beyond where Mavericks usually broke, Alladio and Cahill noticed an odd gray bank on the horizon, like a wall of low-lying clouds or a storm front. It was only when the horizon started feathering at the top, white spray spuming in the air, that they realized: *This is a wave.* Whatever size it was, it dwarfed the sixty- and seventy-footers they'd been dodging all day. There was a split second of terror and confusion and then Alladio motioned desperately to Cahill: they couldn't outrun the wave, so their only hope was to race straight at it and make it over the top before it broke. They managed that, barely, and were rewarded with a fifty-foot free fall on the backside, dropping into the steep trough. Plunging that far on a half-ton machine was as bone-jarring as jumping out a third-story window. But worse, in front of them, bearing down like hell's freight train, was another colossal wave. This one was even bigger.

Again they gunned for the peak, squeaking over the top before the crest started its avalanche, and once again they air-dropped into the trough. But they had to keep going; Alladio could see at least three more waves in the set. By the time they had faced down the last one they were miles offshore, the land behind them obscured by a white scrim of spray.

In the waves' aftermath, Cahill noted to the *San Francisco Chronicle* later, there was an eerie stillness on the water. Oceanic roadkill—dead fish, torn strands of kelp, and broken bits of reef—swirled around them. "Each time we went up [the faces of the waves] I could see all these fissures or ravines in the surface, and there was some kind of crazy light energy vibrating inside the wave," Alladio recalled in the same interview. The whole experience sounded outrageous, like a bad dream or a scene from a disaster movie, something that couldn't possibly have happened in real life. But there were witnesses, and they included veteran Mavericks surfer and documentarian Grant Washburn, who was filming from a nearby cliff when the set broke. Washburn knew these waters inside and out, and he had never seen anything like those waves. He believed they had easily topped one hundred feet.

"The whole hundred-foot-wave thing," I said to McNamara. "What do you think about it?"

"Ah," he said, shaking his head. "We don't have no interest in a hundred-footer." McNamara had an unusual accent, a mix of California dude and Hawaiian heavy. As he spoke, he drew out certain words and clipped off others. "He's got kids, I got kids . . ." He paused so I could take in this mature and cautious stance, and then he delivered the kicker: *"Gotta be 120!"* He and Mamala roared with laughter.

Outside, things still looked nasty. As we approached Half Moon Bay I turned off the highway and into the Harbor View Inn, a two-story motor lodge painted a sickly pale green. Its red neon "vacancy" sign glowed weakly in the mist. I pulled into what appeared to be a Jet Ski dealership but was, in fact, the motel parking lot. Instantly McNamara and Mamala were out of the truck, circulating among the knot of men gathered there, swapping buoy readings and plans of attack. Judging by the crowd, Mavericks had held its own. Jamie Sterling was here, as were big-wave prodigies Greg Long, Mark Healey, and Nathan Fletcher. I saw Dan and Keith Malloy, two of a celebrated trio of brothers from southern California, and a talented pair of Australians, Jamie Mitchell and James "Billy" Watson. Another Australian star, Ross Clarke-Jones, had flown in from Europe, fresh from chasing that continent's storms a few days earlier. John John Florence, a fifteen-year-old wunderkind from Oahu, had come, along with his mother, Alexandra, and his younger brothers Nathan and Ivan. (Note to casting directors: the search for the perfect surfing family need go no further.)

"Who tows John John?" I heard someone ask.

"His mother."

But John John wasn't here to ride. He didn't think he was ready for Mavericks quite yet and had come purely to watch and learn. Hearing this, I was impressed. After all, here was a kid whom nine-time world surfing champion Kelly Slater had pointed to as the future of the north shore, who ripped Pipeline at age eight, and who, at fourteen, had competed in the Triple Crown, one of the sport's most elite contests. In tow surfing there was such a surplus of stories about ill-prepared bumblers getting flung onto giant waves that when someone actually took a smart approach, their actions stood out in sharp relief.

Prickett and Harro arrived and began to unload their gear into a motel room. I circulated in the parking lot, listening to the harbor foghorn's baleful blare, and to conversations that played like endless loops, everyone trying to figure out the day, the wave, the best move, the next move. Nobody knew anything for sure, squinting into the dull light at the unseeable ocean. The Waves of the Gods could be out there, but until the fog lifted it was a closed set. Charging down a seventy-foot face was dangerous enough when you could see it; when you couldn't, it would be safer to drive blindfolded down Highway 1. And Mavericks, they all knew, would be at its craftiest on a west swell. Its currents could change direction, running north rather than south, working against the surfer as he tried to outrun the lip, and then if he fell, dragging him deeper into the impact zone. West swells also made the waves thicker, so when they hit the reef they jacked up like an ambush, tripling and quadrupling in size.

"The fog's gonna lift. In the next hour."

"How long's the drive down to Ghost Tree? Three hours?"

"If the fog doesn't lift, we're gonna go down there. In the next hour we'll make the call."

"Well, hopefully this fog's gonna . . ."

"What we need here is a bit of northwest wind. Clear this shit right out."

"At some point the fog has to lift. Um, doesn't it?"

When the fog was still hunkered down at noon, the riders' agitation levels spiraled. Reports of sunny, sixty-foot Ghost Tree sent two carloads back out on the road, pointing themselves south and hoping some of those waves and a whiff of daylight would still be there upon arrival. Garrett McNamara brushed past the crowd, wearing his wetsuit. Tired of the curbside guessing, he was headed across the road to Mavericks' Jet Ski launch. "I'm gonna go out and take a look," he said. John John, also suited up, followed him.

Prickett emerged from the motel. "Everyone's getting into full panic froth mode," he said. "They're afraid they might get blanked. But we are going out." The camera cases were ready to load onto the boat; we'd wait for McNamara's scouting report and then we'd leave. Prickett, I'd noticed, had the admirable habit of laughing off stress, even as he reckoned with it. Despite his determination to go out in the ornery conditions, he was well aware of how treacherous Mavericks could be. "I was here a couple years ago on a big day," he told me. "I almost died because I got washed through the rocks. I was swimming [shooting from the water] and the next thing I knew these waves were on me. I got *mowed*." He exhaled, remembering. "So we're gonna charge it, but we'll just take a moment to . . . check. We'll hear what Garrett has to say."

"Hey Prickett, your phone is ringing," someone yelled from the room.

He turned to go back inside, shooting a quick glance at the sky. "Is it me, or is it getting lighter?"

Twenty minutes later McNamara returned, storming through the parking lot, his eyes twice their normal size. "GIGANTIC!" he yelled to the people milling around. "You gotta get out there. 'Cause some guys ain't gonna want it pretty soon." The place went into high gear, everyone suddenly preparing to leave. It was, however, still foggy.

Prickett leaned out of his door and motioned for me to come over. "Our boat captain's kind of freaking out," he said. "He doesn't want to do it." He shook his head. "That guy was sketchy from the start. He's a worrier: *Yap yap yap yap yap*." He imitated a frantic chihuahua.

I stared at him. "Can we get another boat?"

"Yeah." Another captain had stepped up, he explained, but wouldn't allow photographers to jump on and off the vessel once we were out there. For Prickett, this wouldn't do; he needed to move around on the water. Therefore he would ride out on a Jet Ski instead. I would carry the extra film gear on board the boat and pass it over the side when required. It was a less-than-perfect plan, but still it was a plan. We headed for the launch.

"NO ONE is getting off the boat when we're out there. Is that clear? I am not covered for people getting on and off the boat. One incident and that's my whole livelihood. This is the most, uh, probably one of the most . . . This is a high-risk adventure. So we are gonna have an extensive safety speech. I need everybody to LISTEN."

The captain was a heavyset guy in a survival suit, pale blond with watery blue eyes. He looked like he wished he had never heard of Mavericks and its tow surfers and its photographers and its sponsors and anyone else who wanted to leave the dock in a thirty-foot west swell in blackout fog. Sweat beaded on his forehead as he spoke, though the air was chilling. All twelve of us—his passengers—and both of his deckhands were bundled in ski jackets and heavy weather gear. I wasn't sure how much we had agreed to pay him to take us out to the wave, but clearly it wasn't enough.

"When we go up over that swell, we will pivot off the back of the boat. You need to STAY OFF THE FRONT OF THE BOAT. You don't, you'll get tossed. And that is not happening today. That is not. Seriously, it is dangerous out there. The wear and tear on my ticker for this kinda stuff . . ."

The guy next to me, a photographer, leaned over and whispered: "When we hit that breakwater, I'm taking bets on who's gonna puke."

The harbor was a wash of gray. Fishing boats bobbed in their slips, their owners having no intention of venturing out in this ugliness. At the docks the water was glassy and still, but that would change in five minutes when we arrived at the harbor mouth. Beyond the long L-shaped break-water, the Pacific was on a rampage. I tucked myself behind a wooden table, braced on all sides. The engines chugged, and we motored out at a crawl. Everyone sat inside the cabin, trying not to look as panicked as the captain. The most self-possessed passenger, by far, was John John Florence, sitting quietly in the corner, a navy hoodie pulled up over his mop of white-blond hair. He stayed out of the nervous chatter.

"This is sick."

"Yeah, this is gnarly."

"I heard it closed out in the channel. Mike got rolled on his Jet Ski, and then they couldn't even *find* it. You can't see shit out there."

"If we end up in fifty-one-degree water, how long do we have? About fifty minutes? Forty-five?"

"Dude, if you're in the water, you're dead."

"You've got to think the captain's done this before. That he knows what he's doing."

"Well, he hasn't done it like this. At low tide, across the reef, with fifty-foot surf."

I looked out the window at the raw cement seascape, the fifty-five-foot boat now starting to buck. A loud roaring noise could be heard in the background. It was an unpleasant combination of effects, made worse a second later when we hit the edge of the swell. The boat reared straight up on a wave, then heeled hard to the side. Camera cases and anything else in the cabin that wasn't strapped down flew into the air. My elbow slammed on the edge of the table. Everyone gasped. We weren't even at the breakwater yet.

The boat idled in place for a moment, as though gathering its bearings before continuing into the barrage. Waves charged us from all directions. "These aren't even the breaking waves," the photographer said. "Wait until you see the breaking waves. We'll have to negotiate those. But if he can't see them . . ."

One of the deckhands stuck his head into the cabin. "Uh, the captain is having some reservations."

"Ah, no way!" A lanky guy who'd been champing at the bit shied back in his seat. "Fucking tour of the harbor!" he said. "That's *bullshit!*"

"Listen," the deckhand said in a stern tone. He looked as though he would very much like to crush the lanky guy like an empty beer can. "People are coming back in because they're getting rolled. Is it worth it? Someone drowns—is it worth it? To see a *wave?*"

"I'm the wrong person to ask," the lanky guy said, crossing his arms defiantly. "I'm a waterman."

Broncoing in the chop, the boat made a U-turn and headed back to its slip. "When you can't see you never know when the big one's coming," the captain said, herding us onto the pier with obvious relief. "The amount of liability involved and the risk of the prevailing conditions— it's too hairball for me."

I stood on the dock, figuring I'd go back to the Jet Ski launch or maybe get up on the cliff, an observation point from where, if the sky cleared, one would see the wave. Lugging Prickett's lead-filled camera case, I began to walk down the dock, stopping often to switch hands. Standing at the stern of his boat, a potbellied fisherman who looked a little like Jerry Garcia scowled at me. He had watched our retreat. "Zero visibility," he said, shaking his head. "Shitty conditions. The ocean's *closed.* Get it? Want to kill yourself? Is that what you want? You. Can't.

Go. Out. Don't you think we would if we could?" He pointed to a boat that was leaving the harbor: "There goes the coast guard." His voice was smug. Somebody was in trouble and he, the Angry Oracle of the Docks, had predicted it.

At the edge of the parking lot the harbormaster's office glowed in the flat, gauzy light. I stepped in to see if I could get a map of the jetty and interrupted an emergency meeting. Three uniformed men were hunched over a nautical chart, looking grave. One of them glanced up with a furrowed brow. "We can't help you right now," he said briskly. "We lost a boat. Now we're trying to find the people."

"I've never been run over by waves this big," Jeff Clark said, tying off his Jet Ski to the launch ramp. "It's the swell direction. As fast as you can go, it's gonna go faster. Pretty radical." Clark wore a fluorescent orange rash guard on top of his flotation vest and wetsuit, a beacon of color in an ocean of gloom. He had returned to shore for a breather, and was surrounded by a local news crew that had been lurking in the socked-in parking lot, hoping for some kind of visual. Their wait was worthwhile: Clark was Mavericks' resident legend.

Growing up within sight of the wave, he began to surf it in the early 1970s despite its heavy roster of dangers; when he couldn't convince anyone else to join him, he paddled out alone. On north swells and west swells and weird jumbled-up swells, frontside and backside, going right and going, outrageously, left, under bright skies and dreary cloud canopies, in the fathomless, haunting waters of Half Moon Bay—for fifteen years Jeff Clark was the only man riding Mavericks. In the early 1990s people finally started paying attention to Clark's entreaties to check out his wave, and by 1994, when Mark Foo jetted over for that fateful swell, Mavericks was no longer a local secret. And the more people learned about the wave's treacheries, the more astonishing Clark's years of solo excursions seemed in retrospect. Even now you'd have a hard time finding anyone who liked the idea of spending a single session out there alone. In a sport where respect is the currency, Clark was a zillionaire.

So when I heard him tell the news crew he had just suffered one of his worst wipeouts ever, I was curious to hear more. *Ever*, for Jeff Clark, was a distinction from within a hefty inventory, thirty-five years of familiarity with Mavericks' bad moods. He leaned against a concrete piling, recalling what happened. Clark spoke with a laid-back California drawl, his words telling a white-knuckle story while his demeanor signaled that however menacing the situation had been, he could handle it.

At fifty-one, his black hair was tinged with silver, but Clark had the powerful physique of a younger man. His eyes, I noticed, were the same ice blue color as a Siberian husky's, a dog known for its single-minded intensity. "It's the nature of this swell," he explained. "It's very dangerous. You could do all the right things and still not make it." The waves, he said, were closing out in a strange way, hooking around at the end of the reef and snapping shut. "It pinches you, like being cut off at the pass. Almost everybody has been caught today." Clark had been squeezed, had to straighten out on a fifty-footer, unpleasant enough, but when his partner, Brazilian Rodrigo Resende, sped in to get him, Clark's glove had slipped off the rescue sled during the pickup and then he was out of time. The next wave in the set, a meaty, nasty affair, not only spun Clark down into the depths but took out Resende and the Jet Ski too.

"It's like a train hitting you, this explosion," Clark said, smiling grimly. "And I'm down. It's so black and violent. I mean you can tell even with your eyes closed, it's *black* black. It is so dark. And then it's not letting me up. And I'm thinking, 'Well, hold out, hold out,' but my limbs are trying to be torn off. I finally got flushed to the surface—*whoosh*—got a breath, and all I could see was another twenty-five feet of whitewater coming. Drilled again." He shook his head with resignation. "It's amazing, you know, sometimes you can actually check out of that kind of abuse to your body. It's like shutting down your computer, logging off. But if you start having two-wave hold-downs, you're playing with the

"It's an animal": Laird Hamilton riding Jaws; Don Shearer flying safety alongside.

(*Above*)
Laird Hamilton and Dave Kalama (right) have been a team since tow surfing's creation in the mid-1990s.

(*Above right*)
"This is the Sport of Kings": Hamilton on the Ski with Darrick Doerner

Gang members: Brett Lickle (left) and Sonny Miller

Hamilton drops in at Jaws, before flotation vests became a lifesaving addition to his gear. Without the vest, a rider's chances of making it back to the surface after a fall are greatly reduced.

The magenta blob: NASA weather maps capture Super Typhoon Nida as it spirals across the Pacific, kicking up giant seas.

The RSS *Discovery* meets the North Sea—on a nice day.

Dr. Penny Holliday aboard the *Discovery* on its ill-fated, but revealing, research cruise in 2000.

Britain's National Oceanography Center at Southampton is one of the world's most acclaimed oceanographic institutes; along with its sister ship, the *James Cook*, the *Discovery* (shown in its berth alongside the building) roams the planet's seas in search of answers.

"Things were getting smashed off": the oil rig Gullfaks C taking heavy abuse in the North Sea.

"A lot of these ships are getting beaten to a pulp": the Iranian oil tanker *Tochal* had its entire bow section torn off by giant waves in the Agulhas Current, off South Africa's southeastern coast.

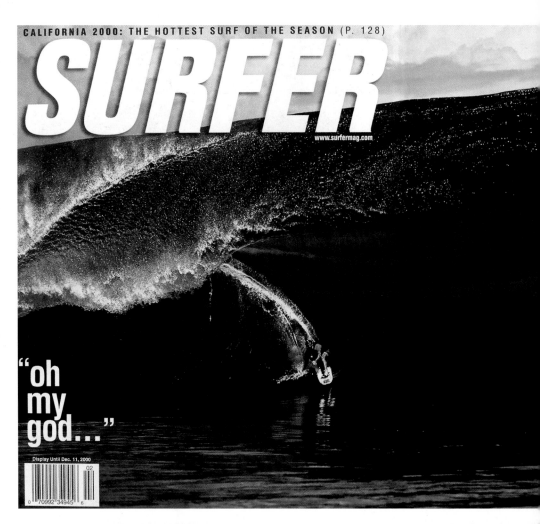

SURFER

www.surfermag.com

"oh
my
god..."

Display Until Dec. 11, 2000

The Tahitian big-wave star Raimana Van Bastolaer

Mesmerized by waves: Jeff Hornbaker

A rare moment on land: ocean filmmaker Mike Prickett

laird hamilton on the heaviest wave ever ridden
exclusive photos p. 104

Game changer: master underwater cinematographer Don King in his element

(Above)
"If he'd fallen he would have been a red stain on the reef": Laird Hamilton makes history in Teahupoo's meat-grinder barrel, shown on the cover of *Surfer* magazine, August 17, 2000.

(Left)
The ride of the day: Ian Walsh drops into Teahupoo on November 1, 2007.

The storm hunter: Harro on the beach

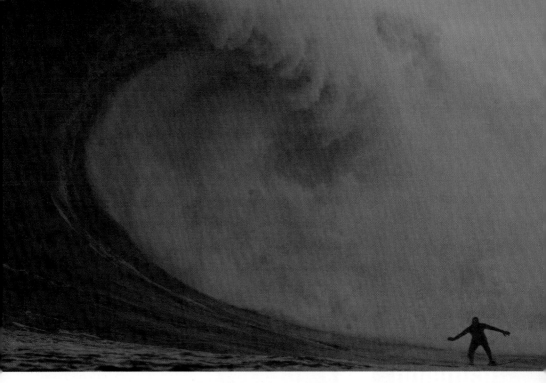

"We've never surfed it this big": Garrett McNamara barely escapes Mavericks' dark jaws on December 4, 2007.

A monster truck of a wave: Australian rider Justen Allport tries to outrun Ghost Tree, off the coast of Pebble Beach, California. Only seconds after this picture was taken the wave's lip broke on top of Allport, snapping his femur into five pieces.

Killers: Brad Gerlach catches a sixty-eight-foot XXL-winning ride at Todos Santos Island, off the coast of Ensenada, Mexico, in 2005.

The moon shot: Mike Parsons rides a seventy-foot monster at the Cortes Bank on January 5, 2008.

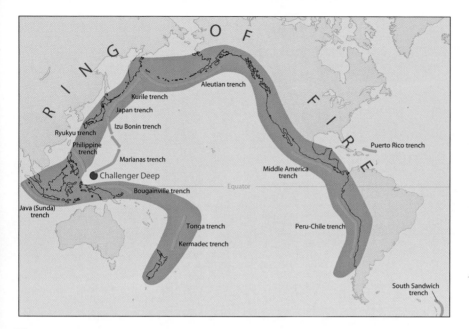

The "Ring of Fire": the Pacific basin's volcano-dotted perimeter, a region responsible for 80 percent of all tsunamis.

(Below left) "It looked as though the ship was heading straight for the white cliffs of Dover": the lordly ocean liner *Queen Elizabeth 2* was hit by a hundred-foot rogue wave on September 11, 1995. Captain Ronald Warwick saw that the wave's crest was level with the bridge (the row of windows between the flags).

(Below right) The Lutine Bell in its place of honor at Lloyd's of London

An artist's depiction of the tsunami that razed Lisbon in 1755, upending life from the Mediterranean to the United Kingdom.

"Ships had simply broken apart like a snapped pencil": a freighter in heavy seas. If giant waves breach the deck hatches, a bulk carrier can sink within minutes.

"And when that violence comes, it is overwhelming": Lituya Bay, Alaska, a haunting place where extreme waves scour the landscape.

The aftermath: in 1958 a 1,740-foot wave scalped the forest around Lituya Bay of trees and soil; it even stripped the trees of bark—with a force exceeding that of a pulp mill. This was only one in a series of epic waves that rampaged across its shores.

Geologist Don Miller documented the 1958 wave's damage, shown here from eye level.

Bad Friday: on March 27, 1964, an earthquake measuring 9.2 rocked the Pacific Northwest, creating tsunami waves that obliterated parts of Alaska, Hawaii, and California. In Anchorage (shown at right), an entire suburb slid into the sea. In Kodiak, Alaska (above), the entire hundred-boat fishing fleet was lost; boats were washed ashore, houses were washed out to sea, and fires raged.

The Wild Coast: marine salvagers rappel from a helicopter to board the *Ikan Tanda*, a foundering Singaporean ship that was being pounded by forty-foot storm waves and gale winds near Scarborough, South Africa. (In the end, the ship could not be saved.)

The hundred-year storm: in December 2009, a massive magenta blob descended on Hawaii. The biggest waves in a century were predicted; harbors and beaches were closed all over the state, but Hamilton (pictured riding Jaws on December 7, 2009), Kalama, and other tow surfers headed directly for the water.

"Let's get a wave": Hamilton and Casey in Jaws' lineup

"Ultimately the objective is to ride the biggest swells the ocean can create": Hamilton on the hydrofoil surfboard, an invention that he believes is the future of big-wave riding.

A stormier, more aquatic future: the world's oceans are increasingly volatile, with average wave heights rising dramatically.

edge." He let out a sharp laugh. "As opposed to . . . playing with the edge."

Along with his own ordeal, Clark described watching Darryl Virostko, a surfer known as Flea, go over the falls on a wave so massive and so deranged that the entire lineup froze when they saw it, fearing the worst. The wave slammed Virostko down smack in the apex of the pit, where the water was at its most frenzied. The fall did not seem survivable. But amazingly Virostko, one of a tight-knit group of elite riders from nearby Santa Cruz, managed to escape more or less unscathed. It was a bit of leniency from a wave that didn't usually bestow such things; the blackness of which Clark spoke was far more representative of the realm.

Clark turned and began to pull on his gloves. "Well, I'm gonna jam," he said, flashing an electric smile. "I'm going back out to get another one."

Right before Clark came ashore I had scouted the cliff at Pillar Point, where I caught a brief and abstract glimpse of what was unfolding in the ocean; for a moment the fog had partially dropped its guard. I saw enormous washes of whitewater that were hard to place into scale, until a darker speck appeared: a Jet Ski. I felt like I was watching a silent movie through a cataract-clouded lens. Mavericks looked towering and brutal, but the distance and the muffled quality of the view made for shadow ferocity. To feel the swell's power you had to be down there. But though I didn't know it as I stood there, wishing I was out on the water, the Angry Oracle had been right: the price of admission today was too high. By midafternoon three people who had ventured into these waves were already dead.

At four-thirty what little light there was in the sky had begun to drain rapidly, northern dusk under a thick cloak. Almost en masse, the riders and photographers returned to the launch. Trailers backed up to the waterline, ready to scoop up the Jet Skis and secure them for the long drive ahead: many of the men planned to travel through the night, chasing the swell as it moved south to Todos Santos, an island twelve miles off the coast of Ensenada, Mexico, to meet the waves at daybreak.

A small crowd had gathered around the ramp, anxious to hear the stories firsthand. There was a sense of relief and triumph and more relief, and while some people seemed wiped out and subdued, others were flying high on remnant adrenaline. McNamara in particular seemed lit from within, wired with energy, his voice raised to a shout. "Gnarliest EVER," he yelled. "I rode one from about a mile out, I don't know how, and I couldn't see anybody for like, at least five hundred yards. And finally, WHOOOOMP!" He laughed maniacally. "Ha ha ha! I love to get pounded!"

Beside me, a smallish guy stood silently amid the hollering and the high fives. Suddenly he turned and said: "I almost died out there today." His face looked tense. Whatever had gone down in the waves, he needed to talk about it.

"What happened?" I asked.

"I came really close," he said in a low, morose voice.

"Were you surfing?"

"I was taking a cameraman out." He shook his head impatiently as though I should know this already. "I lost a Jet Ski and I got caught in a weird place and I took about ten waves in a row on the head. I was stuck over there where they found Mark Foo. I was stuck there, getting pounded, *one wave after another after another.*" He delivered the skeleton outline of his story with a shrill note of panic. "And the fog was in, so I couldn't see. I thought I was going out to sea! And seals were popping up next to me! Yeah, I came really, really close."

A stout man standing on my other side heard this story and leaned in. "Someone did die at Ghost Tree," he said.

"What?!" I said. "Who? When? *Today?*"

"A surfer," he said. "I'm not sure who it is. He drowned."

"Did he hit the rocks?"

"Yeah, I think so. Sad."

Looking around at the launch, it was clear that none of the riders here had received this news. Who had died? And how? In the background I heard McNamara's voice rise above the others': "Yeah, I put him on a giant wave! It was just so perfect and so big and I watched him ride it and he just killed it like crazy. Two big ole snaps. Fucking gnarly bottom turn!"

"It was scary out there," Mamala said. "Biggest barrels I've seen in a long time. Just tall and mutated." He drew out the last word for emphasis.

"We've never surfed it this big," McNamara said. "Very rare."

I saw Prickett standing off to the side, checking something on his camera. He looked spent. "Big black gray muddy water," he said, describing what he'd seen through the lens. "Dark, dark, dark. But there were moments." On the way out there, he told me, the Jet Ski had punched through a wave, flipped upside down, and tossed him into the whitewater with all his gear. "It was a big ordeal," he said. "So we had to get through that."

"Okay boys, come on, let's go!" someone yelled. "We're going right now. We need to get driving!"

"Did you see the wave?" Prickett asked me.

"Sort of," I said. I told him about the aborted boat ride and my visit to the cliff. "Hey," I added, "did you hear about what happened—"

"Well, you'll just have to come down to Todos with us then," Prickett said before I could finish. "It's gonna be just as big. We're on a flight to San Diego at ten."

Just then my phone vibrated. Sean Collins had left me a voice mail. *"Yeah, we had a pretty good day,"* he said in his quiet, low-key way. *"Not fog but mist. And it got big. Fifty-five feet, probably. The only bummer is a guy died here today. His name is Peter Davi."*

Peter Davi was a surfer from Monterey. He was accomplished in big waves, a well-known and much-liked presence on the northern California coast. A third-generation fisherman whose Sicilian grandfather had worked on the Cannery Row of Steinbeck's day, Davi was also a regular on Oahu's north shore, making for Pipeline when the herring weren't running. In that hard-core arena he earned respect from—and a place among—the locals, a group not known for their easy inclusivity. Like the Hawaiians, Davi appreciated elemental things—the beauty of rocks, for instance, or the way morning light glinted on the ocean.

But along with this sensitivity, at six feet three and 265 pounds, Davi—also like the Hawaiians, and for that matter his Sicilian forebears—could be awfully intimidating if the need arose. Yet no one was strong enough to accomplish the task he set for himself when he had shown up at Ghost Tree this morning: rather than towing, he intended to paddle into the waves. On a swell this powerful, that was not only a futile decision but also, it seems, a fatal one.

Pieced together from riders who encountered Davi on the water, a blurry picture of his last moments emerged. After unsuccessfully trying to paddle into waves on his eight-foot gun, Davi sat on the back of his friend Anthony Ruffo's Jet Ski and watched the five-story office buildings roll in. Some of the last words anyone heard him say were "I'm forty-five years old and I want one of those fucking waves." Realizing the only way he was going to get one was by towing in, Davi accepted a ride and surfed what was his final big wave, exiting with a full-face smile. And then he headed in, declining the offer of a lift back to shore. He would get there under his own steam, as he had done countless times before.

But Davi never made it. Somewhere along the way he lost his board, knocked off by the heaving seas, a sneaker wave, or a spasm of whitewater. A spectator glimpsed him swimming near the rocks, but then Davi was swept from sight. Right around that time, Ruffo and his partner, Randy Reyes, had also turned back for shore, motoring in on their Ski. Instead of finding Davi waiting for them, they discovered his body floating near the wharf, facedown in a patch of kelp. Paramedics arrived quickly and tried to revive him, but it was too late. By the time Davi was found he had been dead, they estimated, for twenty minutes.

I soon learned that Davi wasn't the day's only casualty. A crab fishing boat called the *Good Guys* had capsized just outside the harbor mouth—only one hundred yards from where our boat had turned around. The two local fishermen, Benjamin Hannaberg and James Davis, had radioed their intention to come into the harbor, but they never arrived; instead they set off their emergency beacon. The coast guard searched extensively for the men, both in their late fifties, but at the site of *Good Guys'* distress call they found only two scraps of the hull. "A twenty-five-foot fiberglass boat—that's like an eggshell in those conditions," the harbormaster said later. (A week later Hannaberg's body would wash up on shore; Davis's was never found.)

Looking back on the day, Peter Mel, an exceptionally experienced big-wave rider, said the surfers would always remember December 4, "but not for the epic rides, more for the carnage." The low, lightless ceiling, the sight of his friend Flea upside down in the lip of a monster, the loss of Peter Davi and the two fishermen, all of these were images that no one wanted to linger on. "It was about riding to survive," Mel said. "It wasn't about riding to enjoy it. And you could see it on everybody's faces. It was all about 'Get me to the channel, I need to get off this wave as soon as possible.'"

Like Clark and Washburn, Mel lived nearby and had seen deep into Mavericks' bag of tricks. Many times the wave had punished him in the fearsome rock-strewn areas known as the Cauldron, the Pit, and the Boneyard. But on this day even Mel had been floored by the wicked vibe in the water. "It looked like the ocean was folding over itself," he said, describing how the waves rose so sharply that they basically had no backs, while their faces were "like Niagara Falls or something." His voice was somber. "It was one of those swells," he said, "that didn't seem like it was meant to be ridden."

Leaving the launch, I walked slowly to my truck. I heard seagulls, still screaming in the dark, and the steady insistent wind like white noise, and the whine of winches lifting Jet Skis onto land. There were no stars to be seen, only the oily glare of the dock lights. It was hard to imagine that an all-night journey into Mexico lay ahead, but I knew I would go. *"This storm will still be packing a punch,"* Collins had said in his voice mail. *"Todos is gonna be absolutely humongously huge tomorrow morning."* I felt my phone vibrate, and looked down to see a text message from Prickett: *"United 787 to San Diego. 10:15,"* he had written. *"See you there."*

THE GIANT WAVES *WILL* OCCUR IN LITUYA BAY
IN THE FUTURE; THIS POTENTIAL DANGER SHOULD BE
KNOWN TO THOSE WHO ENTER THE BAY.

Don Miller, United States Geological Survey

"I NEVER SAW ANYTHING LIKE IT"

GLACIER BAY NATIONAL PARK, ALASKA

If a person wanted to visit Lituya Bay, a remote fjord slashed into Alaska's west coast just north of Sitka, he would first fly to Juneau. From there he'd take a short flight to the small town of Gustavus, the jumping-off point for Glacier Bay National Park. Next, he'd hire a seaplane. And if he got lucky with conditions, and the usual blanketing fog and drippy weather wasn't parked over the bay, and the winds were resting from their hoedown, and if the pilot wasn't too freaked out to consider the trip to begin with, he would eventually drop down over the majestic snow-capped Fairweather range with its sentinel glaciers, descending over dense wet forests of spruce and alder and cedar and hemlock—the steep hillsides a tangle of living trees, brush, and downed, rotting timber—and then he would see a seven-mile-long, two-mile-wide T-shaped inlet with a small teardrop-shaped island in its center. At first glance, Lituya Bay might possibly (and deceptively) look peaceful. But upon closer examination, as the plane glided down below the treetops, he would notice something startling. A half mile above the water the forest abruptly stops, as though someone had come along with a razor and given it a savage haircut.

During the first half of the twentieth century, geologists puzzled over the strange denuded areas, the visible scars and wounds that pocked the land surrounding Lituya Bay, searching for an explanation. For years they found themselves unable to come to a conclusion. Cataclysmic things had happened in here, everyone agreed, but what kind, exactly— and when? Had a glacial lake burst through an ice dam, spilling into the bay and washing away all the vegetation? Or maybe an avalanche had scoured the land clean. Had there, perhaps, been an epic flood? In Alaska there were many potential sources for the trauma. The region was dotted with active volcanoes, riven by earthquakes, host to landslides and rock-slides and radical conditions of all kinds. For years Lituya Bay was a con-founding mystery. But as the story of its past came into focus and Nature gave some forthright demonstrations of what it was up to, the culprit became clear: giant waves, the largest ever witnessed on earth.

The bay's history was a quilt of stories about wave-induced fear and death passed down by the Tlingit Indians (pronounced KLIN-kit) who'd lived on these shores. According to their lore, entire villages had been wiped out when immense waves roared from the Gilbert and Crillon inlets at the head of the bay (think: the arms on the top of the T). Freak-ish waves had also occurred at the bay's mouth, a narrow, 300-yard-wide passage where a fifteen-knot current collides with the unruly Gulf of Alaska over a shallow bar. The Indians told the story of how eighty men had gone out in ten war canoes and never come back, and of another loss of sixty men in four canoes. They recounted the eerie tale of a native woman picking berries who had returned to her home and found it washed away, her entire clan killed and their bodies draped from trees.

A Russian expedition led by Vitus Bering and Alexei Chirikov inves-tigated the bay in 1741; their scouting boat of eleven sailors rowed from the gulf into Lituya Bay and was never seen again. Bering dispatched another party of four to find out what happened, and it too vanished. At first the Russians assumed that the Tlingits had killed them, but eventually they came to believe that the fifteen men had drowned in the waves. On their heels, in 1786, the French explorer Jean-François de Galaup, Comte de La Pérouse, arrived, calling Lituya Bay "perhaps the most extraordinary place in the world." Three weeks later he ran for the exit, minus twenty-one men and two boats that had been lost to its waters. Before he left, La Pérouse erected a plaque on the lone island in the center of Lituya Bay, naming it Cenotaph Island in memory of the perished seamen. "The fury of the waves in that place left no hope of their return," the explorer wrote. "Nothing remained for us but to quit with speed a country that had proved so fatal."

Throughout the nineteenth century, a stream of boats went down at the entrance, swamped by rogue seas, countless victims lost in the freezing waters—and waves continued to shave the bay's hillsides with regularity. In 1854 a 395-foot wave roared through Lituya with such fierceness that it not only swept away the trees but also stripped them of their bark. History didn't record the wave's human toll, but at that time American and Russian whale and seal hunters often holed up in Lituya Bay (ironically) for shelter, and the Tlingit population that lived on its shores likely numbered in the thousands. Twenty years later, in 1874, an eighty-foot wave rampaged through the bay wreaking more havoc, and then in 1899 a series of huge quakes created a set of two-hundred-footers that cost many of the area's gold prospectors their lives. "We ran from our tents leaving everything behind," one man's account read, describing the panic when the waves surged toward them. Over the decades and centuries there were many wave events, and they all had the same plotline. "Lituya Bay is a paradise always poised just on the edge of violence," one historian wrote. "And when that violence comes, it is overwhelming."

What caused these waves? The Tlingits believed the source was a sea monster named Kah Lituya (Man of Lituya) that lurked in the bay's waters, his lair located deep beneath its pinched mouth. Whenever Kah Lituya was disturbed by interlopers or in any way pissed off, he showed his displeasure by reaching up from below, grasping both sides of the bay, and shaking them—hard. Those who died in the giant waves he created then became his slaves, fated to prowl the surrounding mountainsides as grizzly bears, on the lookout for other humans that Kah Lituya could ensnare in his trap.

As one might expect, the geologists who arrived in the mid-1900s had another take.

Lituya Bay, they concluded, was unique in the world, so perfectly equipped to pump out towering waves that nature might have designed it specifically for that purpose. On its three enclosing sides, steep unstable slopes and sprawling glaciers sheared straight up from sea level to seven thousand feet, loaded with rock and ice payloads that—with minimal encouragement—would go crashing down into the water, creating dramatic, localized tsunamis. (Imagine paving stones being dropped into a bathtub by someone standing on a ladder.)

Nothing in Lituya Bay, however, was minimal or moderate. Instead of gently sloughing avalanches, its surrounding mountainsides convulsed in great wracking seizures brought on by earthquakes along the Fairweather Fault, a jumpy rift that traced the bay's eastern edge (the top of the T). When it came to making megawaves, there was plenty of raw material here: the fault was ideally situated to dislodge large masses of glacier and rock, the mountains had near-vertical faces, and the bay itself was more than seven hundred feet deep. Ravaging earthquakes occurred with startling regularity: between 1899 and 1965 Alaska experienced nine that measured higher than 8 on the Richter scale and at least sixty that measured stronger than 7. In 1899 one big quake punched a section of the Fairweather range forty-seven feet skyward.

During the twentieth century no one was more exposed to Lituya Bay's perils than Jim Huscroft, an Ohio expat who'd come to Juneau in 1913 to work in a gold mine. When the mine closed down in 1917, Huscroft built a cabin on the west side of Cenotaph Island and took up permanent residence there. It was a solitary life but never a lonely one. Huscroft, a friendly man and a spectacular cook, was visited by a steady trickle of mountaineers attempting to climb Mount Fairweather, fishing boats that had anchored in the bay, and the odd enterprising grizzly bear that swam out to the island in search of food. Huscroft raised foxes and brewed beer, fished and grew vegetables and picked berries and baked bread. He built a small landing area for boats. He weathered Lituya's frequent gales, constant fog, and driving rain. He looked at the doleful inscription on La Pérouse's monument—"Reader, whoever thou art, mingle your tears with ours"—and he listened to the sound of ice and rock plunging down at the head of the bay and exploding the waters, knowing as any seasoned Alaskan would that these noises might eventually add up to a more personal kind of danger, that Lituya Bay, his landlord, might one day exact the heaviest payment imaginable for his tenancy. The rent came due on October 27, 1936.

Just before dawn that day Huscroft, then sixty-four, stood at the stove in his long underwear making pancakes, his kitchen light visible to two fishermen, Fritz Frederickson and Nick Larsen, friends of Huscroft's who were anchored just offshore on their forty-foot trawler, the *Mine*. At six-twenty a.m., as Huscroft worked his griddle and the fishermen brewed their morning coffee, a terrible noise began, an overwhelming but toneless din that Huscroft later described as the sound of "a hundred airplanes flying at low altitude." The noise lasted for twenty minutes. Something was going on at the head of the bay, but Huscroft couldn't figure out what; there had been no earthquake to shake anything loose. Aware of the bay's history, he ran outside. Aboard the *Mine*, the two fishermen stood on deck looking anxiously into the distance. There was a sudden quiet as the noise stopped, a sinister, pregnant silence. Then the wave appeared.

Huscroft stared for a moment at the dancing white line swinging toward him like a four-hundred-foot-high sledgehammer—it was maybe four miles away, spanning the entire width of the bay—and then he tore off for higher ground. Larsen and Frederickson lunged to hoist their anchor. Realizing they couldn't escape the wave, they gunned straight toward it, trying to clear the crest. It was more like a wall than a wave, Larsen observed, clutching the wheel as his boat clawed its way up the face. He then realized, to his horror, that the wave's backside was nothing but a sheer vertical drop; the water had drained out of the bay so dramatically that its surface had been sucked below sea level.

Falling into the trough, the two men saw another giant wave hurtling toward them, larger than the first. Then a third, larger than the second. The *Mine* careened on a berserk roller-coaster ride but it survived, as did Larsen and Frederickson. In the aftermath the waves continued, smaller but still forceful, and they banged at the boat from all directions as the bay tried to regain its equilibrium. Huscroft too survived the waves, though not much else on Cenotaph Island did. The trees and vegetation were gone, as were the foxes, topsoil, garden, storage shed, root cellar full of food, dock, most of his supplies, and La Pérouse's memorial. Part of Huscroft's cabin had been washed away, and what remained was severely flooded.

Later, geologists examining shoreline damage and carbon-dating tree rings estimated the waves' height at the head of the bay: 490 feet. (By the time they'd hit Cenotaph Island they hovered around one hundred feet.) In the absence of an earthquake, they guessed that the mechanism for the waves had occurred underwater, an enormous submarine landslide provoked by . . . something. (This theory was never conclusively proved.) For his part, Huscroft never quite recovered from the hit. It was as though the wave had washed his spirit off the island too, and though he continued to live there he never replanted his garden or fully rebuilt his settlement. He died less than three years later.

Life in Lituya Bay went on, the infinite cycle of winds, rains, and storms, the swirling play of the aurora borealis over the peaks, the weightless arcs of gulls and cormorants and auklets, the glaciers silently standing guard. When seas were rambunctious in the Gulf of Alaska, fishermen still dared to take refuge in the bay, nervously crossing the bar at slack tide, darting through the mouth when the waves were quiet. And for a while the waves *were* quiet.

Then in 1958, Kah Lituya went postal.

"Do you know this Mavericks place up by Half Moon Bay? Have you ever run into Grant Washburn? He's a really studious guy. Really knows a lot about waves. And Jeff Clark—whoa! Did it all by himself out there, with sharks and . . . heh, heh, heh." George Plafker, familiar with the local big-wave lore, laughed admiringly at Clark's exploits. He leaned against his desk and crossed his arms. Still strapping at seventy-eight, Plafker, an emeritus geologist with the United States Geological Survey (USGS), was a veteran of the world's most rugged places and one of the foremost experts on nature's worst tantrums. Subduction-zone earthquakes, deformations in the earth's crust, high-speed avalanches, submarine landslides—and the giant waves that resulted from everything on this list—these were all in a day's work. Plafker was especially knowledgeable about Alaska and in particular the area around Lituya Bay. He wore a faded plaid flannel shirt, a fleece vest, jeans, and sturdy boots. Rimless glasses perched on his nose.

I had come to his office in Menlo Park, California, to talk about Lituya Bay's grandest spectacle to date, a 1,740-foot wave that ripped through there on July 9, 1958. Though Plafker himself had been in Guatemala at the time ("Much nicer down there"), his colleague Don Miller, another USGS geologist, had been working close by and was able to survey the bay within twenty-four hours of the event. Together Plafker and Miller had studied Lituya Bay extensively in the 1950s, examining the landscape for clues that would enable them to chart its volatile past. "We speculated a lot about what caused those waves," Plafker recalled. "We knew it was something big, and we had all kinds of mechanisms, all of which proved to be wrong." He chuckled and reached for a thick, dusty pile of folders on a shelf. "It is a unique spot," he said. "Knowing what I know now, I get nervous at the thought of being there. Its history is just: Bang! Bang! Bang!" He handed me an armful of folders.

Among the acres of files, books, and maps in his office, Plafker had stored Miller's original papers and photographs. Miller, who drowned in 1961 while surveying the Kiagna River, north of the Chugach Mountains, had been sent to Alaska to scout for oil reserves. In the course of his work he'd become fascinated by the giant waves in Lituya Bay, generating a trove of research that remained unarchived due to lack of funds. In the heap of material I held was the inside story of a wave the size of ten Niagara Falls. "It should be fairly clear, I think," Plafker said. "We did annotate a lot of this." He leaned over the table and peered at a large black and white photograph, edges curled, that was secured to one folder's cover by an elastic band. "The boats shelter in right behind the bar," he said, pointing at an area south of Cenotaph Island. "When the wave hit, that's where Howard Ulrich was."

In the beginning, July 9, 1958, was a stunning day, noteworthy for its clear skies and crystalline beauty. Just outside Lituya Bay's mouth, several fishing boats had been jolted by an earthquake foreshock, but that wasn't unusual and no one thought much of it. Things were tranquil as evening settled over the bay, though the immediate weather forecast seemed likely to change that. At seven p.m., still daylight in these latitudes, an amphibious plane circled and then touched down on the glassy water. From their beach camp on the bay's northern shore, ten mountain climbers from the Alpine Club of Canada watched its descent. This was their pilot. He hadn't been due until the next morning, but fearing meaner weather he had arrived early to retrieve them after their successful ascent of Mount Fairweather. The climbers began to pack their gear, stashing some of it in what remained of Huscroft's cabin. While they were busy with this, three fishing boats—also worried about a change in conditions—arrived in the bay to anchor for the night.

The boats were of a similar size and vintage, trawlers in the forty-foot range, sturdy as bulldogs and built to withstand the Alaskan seas. The *Badger* was skippered by Bill Swanson and his wife, Vivian; the *Sun-*

more was manned by another couple, Orville and Mickey Wagner. The third vessel was Howard Ulrich's boat, the *Edrie*. Ulrich, who lived just up the coast, knew these waters as well as anyone. With him was his seven-year-old son, Howard Jr. All three boats were part of a close-knit group of salmon fishers working a stretch of ocean known as the Fairweather Grounds. In the Gulf of Alaska dangers arise frequently, and the boats kept in constant touch via two-way radio.

By nine p.m. the three boats were at their anchorages, and the climbers were ready to depart. As their plane lifted off in the rich northern twilight, a curious thing happened. Noisy clouds of birds began to depart the bay too, kittiwakes, gulls, and terns, wheeling frantically as though chased by a squadron of hawks. In their panic to leave, some of the birds smacked into trees and other obstacles, dropping dead to the ground. And at that moment, if you had stood very still and watched the flowers and grasses along the bay's lower elevations, you would have seen that they were trembling.

Standing on deck after dinner, Ulrich noticed groups of Dall's porpoises traveling from the bay out to sea. He saw their dark backs, the flash of their white bellies, moving through the water. The surrounding mountains were hulking, white-capped silhouettes. Anchored in the southern lee of Cenotaph Island Ulrich couldn't see the other boats, though he'd heard their engines. Just before ten, he and his son called it a night.

Perhaps Ulrich was already dreaming of more pleasant things when he felt the first hard tug on his anchor chain. Figuring it had dragged, he ran to the wheel. It was 10:17 p.m., still light enough in the Alaskan summer for Ulrich to see something astonishing at the head of the bay, a vision not even his nightmares could rival: the mountains were *twisting*. "They seemed to be suffering unbearable internal tortures," he recalled later. "Have you ever seen a fifteen-thousand-foot mountain twist and shake and dance?"

Avalanches poured into the bay, 300 million cubic meters of rock and ice plunging thousands of feet to the water. For what he estimated to be two minutes, Ulrich stood rooted on deck, frozen by the scene. "It wasn't fright," he said, "but a kind of stunned amazement." Suddenly an earsplitting crash erupted, and Ulrich saw "a gigantic wall of water, eighteen hundred feet high" engulf the northwestern edge of the bay, ricochet to the east side, and then head directly for Cenotaph Island and the *Edrie*. Like Larsen and Frederickson before him, he tried desperately to raise the anchor, but it seemed to be stuck. Throwing a life jacket on his son, he did the only thing a mariner in his situation could do. He let out all 240 feet of his anchor chain, opened the *Edrie*'s throttle, and headed straight at the wave, yelling a Mayday into the radio: "All hell has broken loose in here! I think we've had it . . . Good-bye."

Plafker showed me into a room next to his office where I could spread Miller's files out on a table. I picked up one with the title "After the '58 Earthquake" penciled on its cover in a neat architectural hand. It contained sheets of 35mm slides and a few faded typewritten pages that looked like an interview of some kind. "From my notes," the first page read. "Diane Olson. F.V. [fishing vessel] *White Light*. Location: About 35 miles from Lituya Bay."

Olson and her husband Ole, it appeared, had been fishing outside the mouth of the bay. Judging from the notes, Diane had total recall, chronicling the events of July 9 down to the minute. Her first inkling of disaster, she believed, happened at 10:22 p.m. "Suddenly it felt as though our boat was being dragged over a corrugated rock." The *White Light* was anchored in sixty feet of water at the time, so that was unlikely. Almost immediately they heard the cacophonous roar of a huge earthquake. "It was then," Olson wrote, "that we turned on the radio."

Panicked, garbled voices cut in on one another, boats all over the area reporting ocean pandemonium. Forty-foot geysers had erupted from crevasses that suddenly appeared along shore; a twenty-foot wave had surged into a harbor near Yakutat; part of an island had dropped one hundred feet into the sea, taking an unknown number of people with it. Underwater cables and oil lines snapped. The reports tumbled in.

Ulrich's Mayday cut through like a siren, silencing the chatter. For several fraught moments the airwaves stayed clear as everyone waited to see if the *Edrie* had survived the wave. After what seemed like an eternity, Ulrich came back on the radio. They had made it, he said, but the bay was a hellish stew of ice chunks, dead animals, and other wreckage, all slamming around in twenty- and thirty-foot waves. "There's big trees, branches, leaves, roots, and everything everywhere I look," he said, his voice cutting through static. "All around me! I've got to get out of here. I never saw anything like it." He paused. "I don't know if I can make it out, but I can't stay here . . . The trees are closing in on me, all around me! We're heading for the entrance."

Everyone feared the bay's mouth would be impassable, clogged with debris and closed off by a rampart of waves, but somehow the Ulrichs made it out. Miraculously, Bill and Vi Swanson also survived. The *Badger* had been picked up by the wave, spun around backward, and hurtled toward the ocean at a height Bill Swanson estimated to be eighty feet above the treetops. Landing hard in the Gulf of Alaska, trees raining down on it, the *Badger* started to sink. The couple was able to scramble into their eight-foot dinghy; they were found two hours later drifting in the dark, shocked and hypothermic. The Wagners weren't so lucky. When the wave appeared, they had run west toward the entrance, rather than facing it head-on. Despite extensive searching, no trace of them was found.

Sorting through a batch of photographs, I came across a ground-level view of what the wave had left in its wake: a mangled battlefield of tree stumps, a forest hacked up and snapped off and strewn everywhere, as though an enormous clear-cutting had been carried out by an army of angry, drunken loggers wielding very rusty tools. For scale perhaps, or because the only possible response in the face of so much bald destruction was morbid humor, Miller had hung his hat on a huge, splintered spruce stump—jagged as glass from having its trunk and branches torn off—and in doing so, he'd inserted a small human gesture into a scene that defied the notion that Nature held any place for us. The picture had such an apocalyptic feel that you expected to see piles of ashes still smoldering, smoke curling from the ground. But the image didn't offer that kind of easy explanation; it had the tomb silence of mysterious disaster. On the back Miller had written a reminder to himself: "Make copy for Howard Ulrich."

It is every geologist's (or tsunami expert's) dream to actually witness the likes of a mile-high wave. This almost never happens, of course; the planet's heaviest spectacles tend to arrive unannounced. But Miller came close to that grail. Sixty miles away, working in Glacier Bay aboard the USGS vessel *Stephen R. Capps*, he felt the shaking and knew this was no garden-variety earthquake. He could do nothing that night, but at first light the next morning he chartered a plane.

Despite the arrival of the promised foul weather, as the pilot circled Miller was able to see the bay while it was still in the throes of its brutal transformation. Rocks fell from the cliffs; water dripped from the land where the wave had hit. Near the top of the T, the bay's surface was sealed in a three-mile ice jam. A huge tongue of the Lituya glacier had broken off in the quake, along with much of Gilbert Inlet's northeastern wall (the left arm of the T). The north side of the bay was plugged with a four-mile raft of trees and other wave-kill, and even more detritus spilled out of its mouth, fanning five miles into the Pacific.

Miller's written accounts of the wave are drily scientific, but even he seemed startled by the abruptly shaved mountainsides, and by an altimeter reading that pegged the wave's uppermost reaches at 1,740 feet. Also, he noticed that the bay's water level had dropped one hundred feet below its usual level. "The bay is a shambles, the destruction is unbelievable," he jotted in his notebook.

Plafker came into the room to see how I was faring, and he looked over my shoulder at a picture of a man in overalls, holding a camera and a notebook and standing amid a snaggle of downed trees. The man was wearing 1950s-era glasses and had a solemn look on his face. "That's Don," Plafker said, smiling at the memory. "Everything around him was just like Pick-Up Stix. And those were *serious* trees. Just . . . gone." He pointed to an image that showed a long stretch of barren rock and ran his index finger across it. "The thing you see here is—there's no soil! It's really hard to imagine how much force it took for the wave to suck it *all* off."

Eventually Miller did those precise calculations. In August and September 1958 he returned to study the bay more extensively. He clomped up mountainsides and measured crevasses and examined tree rings; he calculated water particle velocities and learned that the wave's force had exceeded that of a pulp mill. On Cenotaph Island he noticed that even the limpets, barnacles, and mussels—some of the earth's most tenacious clingers—had been blasted off the rocks. "Not one living shellfish was seen," Miller reported.

In the end, Miller reckoned that the July 9 earthquake along the Fairweather fault—an 8.0 on the Richter scale that was felt as far south as Seattle (where it knocked the needle off a seismograph at the University of Washington)—had shoved Alaska hard: fourteen feet laterally and three feet vertically. It triggered a series of avalanches, which in turn kicked up 1,740 feet of screeching water. The resulting wave, basically an epic splash, had crossed the bay at more than one hundred miles per hour. It razed the landscape for four square miles, trampling every living thing in its path with four fortunate exceptions: the Ulrichs and the Swansons.

The event was harsh. And, Miller realized with a chill, it was likely to happen again.

And again.

Wherever peppy earthquake action meets ocean, there will be giant waves. Which means not only Alaska but the entire North American west coast is poised to produce them. If you zoom out on the satellite map and trace the Pacific basin's outline you're looking at a path that scientists refer to as the Ring of Fire. A majority of the world's active volcanoes reside along this arc, both above water and below. It's the most seismically active place on earth, and the source of 80 percent of all tsunamis. As the Pacific and North American plates—two of the continent-size jigsaw pieces that cover the earth's molten core—grind against each other off California, Oregon, and Washington, those movements create earthquakes. If, in the shaking, the land lurches vertically and enough water is displaced, tsunamis will arise.

Lately geologists have refined their sleuthing methods, taking core samples from seabeds and coastal areas and then searching the layers for odd deposits—seashells crushed in alluvial sand a mile inland, for example, or coral that somehow made its way to the top of a two-hundred-foot bluff. Sometimes they find "ghost forests," places where the trees have been snuffed out by being buried, drowned, torn away, or poisoned by salt. Whenever the ocean leaves these kinds of calling cards, scientists can infer that waves once swept over the area with great force.

Using these techniques, they discovered that a tsunami rivaling the one on December 26, 2004, in Indonesia, had been generated on January 26, 1700, off the Oregon coast, by an estimated 9.0 earthquake. (This was surprising: at the time, the six-hundred-mile-long area that ruptured, known as the Cascadia subduction zone, had been considered kind of sleepy.) No visual description of the tsunami's impact on the Pacific Northwest has survived, but it was likely impressive: the waves' fingerprints show up in the geological record all the way from northern California to Vancouver Island.

Proof that this earthquake spawned a tsunami solved a longstanding mystery: the source of the twenty-foot waves that had walloped six hundred miles of Japan's Pacific coast at that same time, flooding villages and harbors, wrecking boats, killing people, and causing fires that burned down homes. By necessity Japan has always been a tsunami-savvy place—no country is more vulnerable to giant waves—but on this occasion people were caught off guard because they hadn't felt an earthquake. Japanese records describe the day as having "unusual seas" and "high waves." They never dreamed the waves had come from clear across the Pacific.

These days the Cascadia fault is under constant scrutiny. The combination of its location near a crowded coastline (a serious tsunami originating there would definitely hit California) and some recent spooky behavior has scientists worried. There are strong signs that pressure is building on the fault again, and that it's likely to grumble loudly in the not-too-distant future. In 2005 a California Seismic Safety Commission report stated bluntly that "the Cascadia subduction zone will produce the State's [sic] largest tsunami." The report went on to predict "wave heights on the order of 60 feet," warning that building codes were too flimsy, evacuation plans were nonexistent, and people didn't take the threat seriously enough: "Californians are not adequately educated about tsunamis and the risks they pose." As if to illustrate their point, on June 14, 2005, a 7.2 quake rattled the seafloor near the Cascadia fault; when the tsunami warning went out, it was widely ignored. It was due only to luck that the earthquake hadn't occurred in the subduction zone itself, and no waves arrived.

If any West Coast town should be unnerved about all this, it is Crescent City, California. Located just fifteen miles south of the Oregon border, Crescent City—named for the sweeping bay that fronts it—is ideally situated to bear the full brunt of a Pacific tsunami. It faces the direction from which the waves would likely be coming, and there are no offshore land masses to shelter it. On the contrary, a nearby seamount coupled with a shoaling ocean floor creates the perfect bottom contours to focus a wave's power—in much the same way as Jaws' fan-shaped reef and Mavericks' underwater shelves do. Downtown Crescent City sits smack in the tsunami-impact zone, a fact that became tragically clear on March 27, 1964—Good Friday, ironically—when a magnitude 9.2 earthquake near Prince William Sound, Alaska, knocked the planet back on its heels, causing water as far away as the Great Lakes to slosh around like a roiled bathtub.

The earthquake's impact on Alaska reads like a list of special effects for a high-budget disaster movie: gaping cracks opened in the ground releasing clouds of sulfurous gas; areas of land suddenly liquefied. Anchorage was all but destroyed that night; an entire suburb slid into the sea. The port city of Valdez was assailed by fifty-foot waves and ended up partly underwater, and in Whittier, population seventy, a pair of forty-footers killed thirteen. At Seward, an oil-storage depot exploded into a fireball, and giant waves picked up an oil tanker and deposited it on land. The waves, now filled with flaming debris, went on to hit the Texaco oil installation, and it too exploded. Fiery forty-foot walls of water wiped out Seward's waterfront, its power plant, and most of its houses. These fire-waves then struck the railyard, where they swept a 120-ton locomotive with an eighty-boxcar train more than three hundred feet inland. The boxcars, also filled with oil, burst like popcorn. Meanwhile the fishing town of Kodiak lost its entire hundred-boat fleet.

The waves sped south toward Canada, smacking Vancouver Island, and continued on to Washington and Oregon. In all of these places they caused destruction and death, but on a mercifully smaller scale. Californians had received warnings that the tsunami was headed their way, but no one was overly concerned. The waves seemed to be fading.

Until they arrived at Crescent City.

High tide had risen and it was close to midnight when the Three Sisters showed up, a trio of waves surging south under a starry, full-moon sky. These first three were ocean Valkyries; they leveled the lower part of Crescent City, scouring two miles inland. Power lines collapsed, fires erupted, people were pinned against ceilings in flooded buildings. Twenty-nine blocks were left underwater, 172 businesses and 91 homes erased. Ten died. But it was the fourth wave that delivered the knockout punch, winding up by draining the harbor, and then rushing back at the land, coming in as a malignant black wall studded with logs, metal, plastic, glass, cars, trucks, home appliances, junk, treasures, bodies.

It was a very bad night. Entire buildings were knocked off their foundations and dragged away. More things exploded. A house ended up on Highway 101. And water, everywhere there was water, swirling like the contents of a demonic blender. The world as everyone in Crescent City knew it had turned darkly aquatic.

"You know, there's no way anybody really observes giant waves, tsunami waves," Plafker said as we gathered up the files. "First of all, you're scared to death after a big earthquake. And then you can't see them because usually they're breaking far offshore. Or you're running like heck. Heh heh heh."

From the photos I'd seen and the notes I'd read and the stories I'd heard about the kinds of waves that Plafker and Miller dealt in, the idea of running seemed quite inadequate. Who, for God's sake, can visualize a 1,740-foot wave in the furthest reaches of their imagination? (Besides Howard Ulrich.) Even Hamilton might have a hard time with that one, I thought.

Earlier in the year Garrett McNamara and Kealii Mamala had attempted to surf the waves made by a calving glacier near Cordova, Alaska. The adventure had sounded good in theory, and they arrived full of bluster. "We like to do new things," McNamara told me. But then a chunk of ice the size of Yankee Stadium had plunged four hundred feet off the glacier, and the thirty-five-degree water had exploded into an unpredictable, confused mess of a giant wave, and McNamara, sitting nearby on a Jet Ski, realized: "I was freaked out." Not even seventy-foot Mavericks or eighty-foot Jaws or max-Teahupoo could prepare a rider for nature's superheavyweight round. "I couldn't . . . I was mind-boggled," McNamara recalled, uncharacteristically tongue-tied. Though he and Mamala did manage several rides, McNamara remained on edge. "I was very concerned to say the least. I was no longer there for my passion, or the rush. I just wanted to get the hell out."

THIS IS NOT YOUR FATHER'S BAJA.

Surfer Magazine

KILLERS

TODOS SANTOS ISLAND,
MEXICO

I'll tell you, I'm more freaked out about the bandito thing than anything else," Tony Harrington said. "It's heavy. We probably shouldn't be going at night." We were standing in front of the San Diego airport beside a mountain of camera gear and surfboards, waiting to load them into three trucks. It was midnight. From here we were headed for the border, and then south to Ensenada. Harro's eyes were bloodshot; he hadn't slept in more than forty-eight hours, and he'd spent all day on the water at Mavericks. The previous day, on Oahu, he had been caught inside by a thirty-foot wave, unable to escape, and his Jet Ski had gone pinwheeling in the whitewater with him on it, a forty-pound camera housing strapped around his neck. The wave's impact had clanged Harro and the Jet Ski together like a pair of cymbals, and now the extent of the damage was clear: his entire groin was bruised "black as the ace of spades."

"Uh, what 'bandito thing'?" I said.

The police in Baja, Harro explained, or banditos impersonating the police, or most likely both, had been on a tear for the past few months, targeting vehicles carrying surfboards. Anyone displaying thousand-dollar toys, the banditos' reasoning seemed to go, would be a rich vein in a shakedown. For surfers, hopping over from southern California to ride

Baja's uncrowded Pacific breaks had long been a popular jaunt, considered safe unless you did something outright stupid, like flash a wad of cash in a Tijuana bar.

A couple of years earlier I had spent a month camping and surfing all over the Baja peninsula with two friends. We slept on dunes overlooking the ocean and had amazing waves to ourselves everywhere we went. When a spooky set of headlights advanced on us one night at a remote campsite, it wasn't a bandito raid but rather the UC Davis Botany Club looking for a place to light a really big bonfire. The only trouble we ran into during the trip was difficulty finding someone to fix our transmission on Christmas Eve (though eventually we did). We drove with our surfboards plainly visible on the roof, and in anticipation of official demands for petty bribes we carried a small cooler full of Monterey jack cheese, which seasoned Baja surfers had told us was an acceptable substitute for cash.

That era of innocence was over apparently, and now Baja was a snakeball of drugs, guns, corruption, and general lawlessness. Two weeks earlier a body had been hung from a highway overpass only a stone's throw from the border. Decapitation appeared to be the regional specialty; heads were found in trash cans and car trunks and, in one brazen instance, the middle of a nightclub dance floor. Only the previous week at Rosarita Beach, a mangy resort just south of Tijuana, gunmen had stormed the police station in an attempt to assassinate the police chief, killing one of his bodyguards (you know you're in Mexico when the police chief needs a bodyguard) and injuring four others. The U.S. State Department warned of "increasing violence," adding that "criminals, armed with an impressive array of weapons, know there is little chance they will be caught and punished. General public should avoid all travel to the Tijuana and Ensenada region for the next 30 days."

Getting specific, the *San Diego Tribune* warned: "Surfers have reason to be especially wary about venturing into Baja California after a spate of armed robberies by paramilitary style criminals." On at least six occasions since Labor Day, surfers had been yanked out of their cars, tents, or RVs by men in ski masks brandishing semiautomatics; their vehicles and everything they contained, taken. Victims had also been handcuffed, kicked over steep embankments, assaulted, raped, abducted, and shot.

And, well, here we were—about to take a nocturnal drive through the bandito hot zone, three rented trucks packed with state-of-the-art photo equipment for a feature film that Prickett was shooting about a Mexican surfer named Coco Nogales, all of Harro's gear, plus that of another photographer, Larry Haynes. Also traveling with us were the Australian surfers Jamie Mitchell and James "Billy" Watson, with their quiver of boards. In essence, we were a flag-waving bandito motherlode.

As we prepared to leave, Prickett gathered everyone on the sidewalk. "Okay, so we stay together," he said. "If they pull one of us over, we all get out. We can give them all our money at once and get out of there fast." Settled on that plan, we jumped into the trucks. Our first stop, the border crossing, did not inspire confidence when a Mexican official leaned through the window wearing a ski mask. But he waved us on, and we passed through the outskirts of Tijuana, land of cheap tequila, cut-rate prescription drugs, and any other brand of trouble one cared to sample. Our headlights illuminated hard-faced men standing around barbed-wire fences and dusty storefronts, someone lying on the ground, the shadowy silhouette of a uniformed *policia* on a horse.

At first the night was wide open and starry, but then the same smothering fog that had plagued us in Half Moon Bay began to slink in from the sea, as though it were escorting the swell south. The scenery blurred and we slowed to a crawl, which freaked everyone out, not only because people were tired to the point of hallucination but because at this speed we were surely bandito bait. "No Maltrate Los Señales" (Don't Mistreat Signs) read a sign that was riddled with bullet holes. At one point we came upon a car parked sideways on the highway as though to block it off, but there was no one inside. Near Salsipuedes, site of a wave that

The Surfer's Journal called "the thickest, meanest point break on the west coast," we stopped at a mist-ghosted Pemex station manned by an attendant who was missing an eye, and we stocked up on bottled water and granola bars for the boat. By the time we arrived at our hotel, it was three o'clock in the morning.

Should *The Shining* ever be remade with a Mexican theme, the El Coral Hotel in Ensenada would be perfect for the set. It was cavernous and deserted and our footsteps echoed on saltillo tile. My ground-floor room, located in a wing so isolated that the electricity had to be turned on specifically for me, opened onto a patio ringed with yellow crime scene tape. I was exhausted enough that I might not have noticed except someone had left the sliding-glass doors wide open, and the wind rushed through the room, causing the drapes to billow. Trying to close everything up, I discovered that all the locks had been torn off the doors, leaving splintered holes. By the time I trudged the half mile back to the front desk to switch rooms, moved, discovered that the locks had been torn off the doors in my new room too, and decided I was too tired to care anymore—it was ninety minutes before we had to leave for the harbor.

"The Mexican Coast Guard has closed the harbor," Prickett said. "The military is standing guard. They're not letting anyone out." We stood outside the lobby in the pitch-black predawn, wind gusting and palm leaves tossing. Obviously the swell had arrived.

"So we're on hold," I said.

"No, we're going out." He looked unconcerned.

"But you just said the harbor was closed. That the soldiers—"

"Yeah. That's why we're leaving from somewhere else."

The gear-loading ritual commenced, the photographers moving methodically, without benefit of fresh, rested brain cells. After thirty-six hours of wave-chasing, I felt cotton-headed and nauseous and like a hand was squeezing my kidneys, but I could hardly complain: Prickett and the others were closing in on sixty hours of sleeplessness, a few catnaps at most, and were preparing to spend another day facing off against waves so intimidating that boats weren't supposed to be out in them, let alone surfboards. Others had it even tougher. Anyone who needed to tow a Jet Ski here had spent the last twelve hours behind the wheel, driving down the coast from Ghost Tree or Mavericks.

I'd noticed early in this project that big waves and extreme behavior went hand in hand, and I knew, anecdotally anyway, that dropping onto the likes of a sixty-foot face was a sensation so potent that nothing else—sleeping or eating, for instance—could compete for a rider's attention. I was constantly asking them to translate that experience into words, not only the men I'd witnessed in action but the greats from previous eras.

Ricky Grigg, a champion big-wave rider in the 1960s who went on to become an eminent marine scientist, told me that the risks of riding giant waves were dwarfed by the reward: "Ecstasy beyond words. Mentally, physically, spiritually, it's the highest place I can imagine being." Feeling oneself connected to the ocean at the apex of its power, Grigg emphasized, was utterly addictive. "You keep pushing the envelope more and more, and the curiosity and the thrill of it get all tangled together," he said. "That's why these guys are so driven."

Grigg spoke poetically about waves, but the analytical side of his brain was never far behind. He believed that a person drawn to this force was genetically predisposed, with a "savage DNA still remnant in his heart and mind." This kind of surfer was about as likely to spend his time slumbering when giant waves were on tap as a hunter stalking his prey or a soldier in the middle of a siege. No one could live with this intensity all the time, of course, but when a major swell came along, it was as though a switch got flipped. While the average person who stayed up for three days walking a physical knife-edge would blow a gasket, for someone like Garrett McNamara it was a hit of pure oxygen. "These guys have two personalities," Grigg explained. "One's gentle, the other's a madman."

Grigg's friend and contemporary Greg Noll, a larger-than-life legend who rewrote the rules of big-wave surfing in the pre-tow era, described the feeling in more physical terms: "That rush! I can't explain it," he said. "When you blow down the side of a wave and the thing's growling at you and snorting and all that power and fury and you don't know whether you're gonna be alive ten seconds from now or not, it's as heavy an experience as sex! If you surf, you know. And all the rest of you poor sons of bitches, I feel sorry for you." Far from striving for a balanced lifestyle in pursuit of his sport, Noll espoused whatever extremes it took to get the ride: "I would have gotten myself shot out of an elephant's ass if it would get me on a bigger wave."

Hamilton, who was not usually at a loss for words, hadn't said much about what it was like to ride a seventy-foot wave, though I'd asked him many times. It wasn't that he didn't want to tell me, but for him the experience defied analogy. Verbalizing what it was like to ride Jaws, he said, was like "trying to describe a color in words." Another time I had asked the question he answered without speaking, by just touching his heart. For Hamilton these waves were less a singular experience than a basic need, up there with breathing.

The passion that big-wave surfers feel for their sport can be chalked up to a number of factors. Consider, for instance, the potent cocktail of neuropeptides the body manufactures when faced with high-intensity situations such as falling in love or narrowly escaping disaster. These chemicals, which include endorphins (the substance responsible for "runner's high") and oxytocin (known as the "cuddle hormone" for its blissful effects), are natural opiates, biological relatives of heroin and morphine. That is to say, one can easily get hooked. And it's not like the person who wants to ride a hundred-foot wave is looking for low-grade stimulation to begin with.

Add to this the scarcity of their quarry—surfable giants—and I could understand the level of obsession I was witnessing on this trip. Sleeping away a monster swell? The real madness for a big-wave rider lay not in pushing the limits of his endurance but in missing the chance to do so. As Jamie Mitchell and James "Billy" Watson, the two surfers who'd traveled down with us, loaded their boards into the truck yet again, I could see the excited energy that surfers refer to as "stoke" on their faces and in their movements. They showed no sign of wanting to be anywhere other than exactly where they were, shoving off for Killers from this Mexican Bates Motel. Leaving it in the rearview mirror, we drove away.

The marina smelled like fish, and not in a fresh sushi kind of way. A sad little boardwalk fronted the docks, its scruffy *tiendas* and nacho shacks shut tight until a more reasonable hour. *"Prohibido alimentar a los lobos marinos,"* a sign warned, adding helpfully in English: "Prohibited to feed to marine wolves." Watching the production crew for Prickett's film organizing a breakfast buffet on our boat, laying out platters of sticky buns and a large frosted cake, I wondered if anyone had told them about the twenty-five-foot swell we'd be encountering. Personally, the only thing I was planning to ingest today was Dramamine.

The Mexican Coast Guard presumably occupied elsewhere, we boarded the fifty-foot cabin cruiser that would take us twelve miles offshore to Todos Santos, a one-square-mile island that served as a perch for a decrepit lighthouse, two rare species of snake, and not much else. The waters around the island, however, were anything but dull. Off its northwestern tip, a craggy underwater canyon funneled the swell energy toward a gauntlet of rocks (every big wave needs its impaling obstacles) at the foot of a cliff, creating a break known as Killers. Tall and pretty, brutal despite its lovely appearance, Killers had dispensed some of the largest waves ridden in North America, faces in the seventy-foot range.

Overlooking the docks, a billboard-size Mexican flag waved lazily in the wind. Our boat steered slowly out of the marina and into a lumpy and disorganized sea, but it was three or four miles before we could feel the rolling, straining energy that signaled a long-period swell. Sean Collins had warned that conditions would peak early; we could only hope that we weren't too late. (To the surfers' frustration, it wasn't unusual for the best waves to arrive in the darkness. During the winter, obviously, there was a fifty-fifty chance of this.) Unlike yesterday, the weather looked promising. The sky cast a silvery shine on the water, and overhead the clouds hung soft and hazy, layered thinly enough to hint at sun.

I sat on a cooler on the back deck and listened to Prickett and Mitchell debate their favorite waves. Oregon's Nelscott Reef, Mitchell declared, was "super-rippable." This wasn't much of a distinction, though, because for Mitchell most waves fell into that category. Like many Australians, he had never met a water sport he didn't like: along with his expertise as a competitive swimmer, canoeist, standup paddle surfer, big-wave surfer, and tow surfer, he was the eight-time winner of Hawaii's prestigious Molokai Channel paddle race, a five-hour, thirty-two-mile gruel-athon in which competitors kneeling on sixteen-foot, specially designed surfboards stroked their way between Molokai and Oahu. Blond, rugged, and wildly accomplished at thirty, Mitchell had so much going on that you could almost forgive him a bit of arrogance, but there wasn't any.

Prickett, like others in the big-wave caravan, had hunted down swells all over the globe. He'd filmed in the jewel-colored lagoons of Tahiti, Fiji, and Indonesia and in the waters off South Africa, murky as a haunted basement. There he had run into a great white shark that he estimated to be twenty feet long: "A VW bus is what it looked like." *Thank God*, he thought when he saw it, for the electronic shark-repellent device he wore on his ankle. But the shark kept coming closer, showing no signs of repulsion, so Prickett, getting nervous, glanced down at his ankle for reassurance. The battery light was off.

It was a story he could joke about now—the shark, after all, hadn't attacked him—but there had been other brushes with peril where things hadn't worked out so happily in the waves. Like the time at Pipeline when he'd been bashed onto the reef and "stuffed into a cave," his leg pinned between two rocks. As he struggled to free himself, his left knee was wrenched from its socket, and every ligament shredded. A Jet Ski zoomed to the rescue but before Prickett could break loose he took a five-wave set on the head, all with his kneecap somewhere on the back of his leg. Nonetheless, his favorite place for waves, he claimed, was still Oahu's north shore.

As we neared Todos Santos, the water perked up and the sky became serious. The pretty pastel light gave way to a cranky gray horizon, partially obscured by clumps of fog. The sea surface, a dull green inshore, deepened to a rich navy blue, dark and luminous at the same time. The boat traced the undulating troughs and peaks of the swell, the type of roller-coaster motion that makes people suddenly lunge for the railings. Brief gusts of an offshore wind blew spray in the air.

At most giant wave breaks, the ideal wind is none at all, though that rarely happens. The second-best scenario is a light offshore wind—breezes pushing directly into the wave's face, making it stand a little taller. The worst thing, a guaranteed surf destroyer, is an onshore wind coming from behind the break and shoving the water forward. This results in a sloppy crumble, a type of wave that surfers call a mushburger, and along with being aesthetically unpleasing it can also be dangerous. There are several places on a big wave where the rider has a split second to make a critical move; the most important one being the drop, the jump over the peak and onto the face. Botching the drop can result in being swept over the falls, a place nobody wants to be. Where the wave's lip hits the water, it cleaves open the surface like an ax; to be caught in that spot is effectively to plunge down an elevator shaft. Not to mention that if the lip itself somehow lands on a rider, the result can be anything from a broken neck or femur—both have been known to happen—to death. An unfavorable wind heightens these risks. It means the wave will have an ill-defined peak, an unstable take-off spot from which to attempt the drop. Imagine a skier trying to get traction in an avalanche, or a long jumper digging his toes into quicksand.

I heard the engines slow and went to the bow to scout our surroundings. The captain maneuvered slowly into the channel, trying to find that tricky balance of good angle + closest position + avoiding disaster. Along with the wave, he had to account for wind, tide, currents, and other boats. Killers was known to shift direction, sixty-foot waves suddenly rearing up in unexpected places. We chugged into a circle of boats and cut the motor. The captain was excited. *"¡Olas grandes!"* he yelled, pointing at the break. "Tsunamis!"

Sure enough, there were whitewater explosions in front of us and a dozen tow teams zipping around, but despite the captain's enthusiasm the waves were disappointing. They were smallish, in the twenty- to thirty-foot range, and from the boat, anyway, they looked crummy. "This is shit!" Prickett said, frowning. "I hope it hasn't peaked already. Maybe it's still coming up." Mitchell, Watson, and Harro voiced optimism that conditions would improve. Wherever there were big waves—or the potential that maybe, later, somehow, somewhere, there would be big waves—I expected to hear an endless stream of weather-related hindsight, conjecture, predictions, and musings. It seemed the men often vocalized what they *hoped* would come true, as though saying something out loud might nudge it closer to happening, less observation than incantation.

"I think it's still building. For sure."

"Hmm, well that was a cleanup set. It's coming up, all right."

"What's with this fucking fog? It's got to lift."

"This stuff? This stuff will lift for sure."

Prickett stood at the stern, hands on hips, surveying the water. "Oh, there's Snips and Gerr," he said. This was big-wave argot for the team of Mike Parsons and Brad Gerlach, who had ridden at Ghost Tree the previous day. Former competitors on the professional surfing circuit—archrivals, to be exact—the duo had earned their way into the highest echelons of respect. Parsons and Gerlach were early tow-surfing converts, trading competitions in smaller waves for roaming the world in search of the giants. They were among the first men to ride the Cortes Bank, and they regularly scouted the west coast of North and South

America, from Washington clear down to Chile, for new and untrammeled waves. Parsons and Gerlach lived in southern California and had cut their big-wave teeth at the foot of this island. But Todos Santos was home base. If anyone could figure out what the ocean was up to on this odd, mercurial day, it was these two.

Killers wasn't the world's most fearsome wave. It didn't have Teahupoo's cantaloupe-hitting-the-pavement fear factor or the Mack-truck-on-steroids power of Jaws or Mavericks' devilish soul. Regardless, no wave was to be trifled with when it rose to the size of a middling office tower, and on the right day Killers could punch above its weight. Early in his career Parsons had broken his nose, blown out his knee, and endured a wicked hold-down here, a wipeout that after ten years he still rated as his worst ever. And in 2005, in almost identical conditions to today's, Gerlach had ridden a wave here that measured sixty-eight feet tall.

They sat on their Jet Ski, assessing the waves and the crowds. Clearly, it was amateur hour out there. I saw one tow team that consisted of a gray-haired man and a kid who looked to be about thirteen. The waves were simply too small for towing to be the honorable approach; the only people who would tow onto a twenty-foot wave were those who shouldn't have been towing at all. In tow surfing's earliest years there was much sniping about its legitimacy, mainly because people guessed, correctly, that scenes like this would result.

Used to be, it took years for a surfer to build the experience and paddling skills he needed to launch himself into a big wave. Now anybody with a credit card and a partner could do it. But that didn't mean he should. "The best surfers are the slowest ones to tow," Hamilton had said. "After boot camp you don't immediately become a general. Unless you can ride thirty-foot surf paddling in on your stomach, there is no way in the world you should be behind a Jet Ski." The corollary was that, if a wave could feasibly be paddled into, and there were riders trying to do that, then the Jet Skis stayed in the channel. Tow surfing, Hamilton emphasized, had been created for one purpose only: "Waves that would otherwise go unridden by the best surfers in the world."

Those were not the waves that were breaking at Todos at this moment. With the arrival of Mitchell, Watson, Greg Long, Rusty Long, Jamie Sterling, and Mark Healey, some of the most accomplished paddle surfers around, the father-and-son Jet Ski duos were about to be side-lined. If Killers hit fifty feet, then they'd have a discussion; in the mean-time the waves would be caught by humans rather than by machines. Mitchell and Watson grabbed their guns, long sleek boards with pointed noses, made specifically for paddling into big waves. Prickett, meanwhile, decided that conditions were manageable enough to swim, to tread water on the wave's shoulder holding his forty-pound camera housing, ducking underwater when catastrophe threatened. He pulled on his wetsuit and fins and disappeared over the edge.

Gazing at the break, I was underwhelmed. Killers didn't seem very killer-esque. But as I climbed onto the bow and sat down to watch, a forty-foot set came in and suddenly the wave revealed a less easygoing nature. The face stood up tall, and in its center a large circular boil appeared like a trapdoor. Caught inside the cascading peak, surfers dumped their boards and dove off to the side. The lucky ones punched through to the back and were able to recover quickly; the unlucky found themselves washed into the spiky rock field. Just a little wakeup call to remind everyone: the ocean could deliver a stiff backhand any damn time it pleased.

Throughout the morning the swell lurched in, coughing up hair-balls and gems with an emphasis on the former. In the water, people were frustrated. The waves hovered in the fifty-foot range, right on the border between towing and paddling, and because of this, most of the best waves were empty. The paddle surfers couldn't quite catch them, although Greg Long and Jamie Mitchell both scored impressive rides. As long as the pad-dlers were out there trying, however, the Jet Skis couldn't swoop in. I saw McNamara and Mamala drive by, looking bored. Sitting on the deck watching failed attempts to catch a wave, I heard a bitter voice from a nearby Ski: "Well, well. There goes another paddle surfer not making his wave. *What* a surprise."

It was a day that couldn't make up its mind. The waves were formidable; then they were not. The sun would come out, only to disappear. The temperature alternated between bathing-suit hot and ski-jacket cold. The fog lifted and then oozed back in. After a few hours of this, Prickett swam back to the boat. Hoisting himself over the gunwale, he held up a battered object the size and shape of a large telephone book, mummified in black gaffer tape. "Look what I found," he said, holding it up and laughing. Clearly it was an orphaned brick of something illegal. Prickett described how, while swimming, he'd felt a hard object bump him. (In the ocean, this is never a comforting sensation.) He batted whatever it was away and kept going, nervously. "So then I get hit again!" This time he saw the object suspended near the surface and scooped it up out of curiosity. We gathered on the stern to examine it.

"Cut it open!" someone yelled. "Yeah! Let's see what it is!" A box cutter was located and Prickett sliced open his contraband. Inside: a soggy, tight-packed mess of marijuana. We were all bent over it, discussing whether it was still smokable, when the captain, wondering what we were doing, wandered back.

When he saw the ten-pound lump of weed, his eyes bugged and his mustache quavered with fear.

"*¡No en el barco!*" he shouted, making a frantic hurling gesture toward the water. Get it off the boat! "*¡Peligro! Ay!*" He clutched his head. If the drugs so much as touched the deck, he said, Mexican customs would seize his boat.

We took some photos of it and then quickly set it adrift, spilling open and primed to become fish food. Prickett watched it float off. "I wonder what the story is," he said. "How it all played out. Was it a drug deal gone bad? A smuggler's boat that sank?"

"I guess we'll never know," I said. "But you do wonder what else might be out there bobbing around."

Prickett ate a quick lunch and then prepared to head back out. "What's the water like?" I asked. He thought for a moment. "It feels angry," he said. "There's a current and a lot of chop. You can tell the waves have come from a huge storm." The filming was difficult, he added, because of the extreme peaks and troughs. "When you fall down in a hole, it's hard to make a judgment call about where to swim to get your shot." He said this cheerfully, and then he jumped over the rail and swam off.

Afternoon commenced. I sat on the bow, sheltered from the wind and with a direct sight line to the wave. The sets became more sporadic, good waves few and far between, and I began to feel drowsy. The swell appeared to be fading, nothing dramatic had happened in hours, and I drifted into hazy inattention. Out of the wind, the warm sun sparkled on the water. Seagulls and frigate birds soared overhead. It was all very picturesque if you ignored the discarded plastic bags, assorted garbage, and jettisoned narcotics swirling in the currents.

Then out of the blue, a horn sounded, blaring an emergency warning to everyone in the channel: massive set on the horizon. There was a sudden scramble, boats motoring up in case they had to flee, surfers frantically clawing their way toward the incoming waves to avoid getting caught inside, Jet Skis roaring to life. I jolted upright, staring—and I could not believe what I saw.

A wave unlike the others had arrived: a true freak. It was the result of God knows what trickster energies in the Pacific, an eagle rampaging through a parade of chickens. Instinctively I flinched as it rose into a sheer cliff, flicking the surfers off its face with casual violence. It was the biggest wave I had ever seen, later estimated to be on par with Gerlach's sixty-eight-footer, and watching it I felt amazement and fear and humility, and through that prism of emotions, I recalled something Hamilton had said: *"If you can look at one of these waves and you don't believe that there's something greater than we are, then you've got some serious analyzing to do and you should go sit under a tree for a very long time."*

By this point I had seen plenty of waves in the fifty-foot range, and though they were truly impressive, until now I hadn't felt the kind of awe that this wave inspired. Because, I now knew, when a wave grows beyond sixty feet tall, it does something *different.* As the wave stood up to its full towering height it hung there, poised on the brink, and instead of immediately beginning to break, the lip plunging over the face and expelling the energy, it advanced as a vertical wall. It was the ocean's ultimate threat, and so the ocean let it hang out and show off and strut for an extra few beats, its crest feathered with white spray and its face booby-trapped with boils, bumps, and turbulent eddies. And as the wave hung in the sky, suspended between beauty and fury, those seconds stretched like elastic, like a terrible void into which all things could be swallowed up forever.

When it finally did break, that too seemed to happen in slow motion, the whitewater rumbling toward the cliff. Perhaps the human brain becomes overwhelmed when trying to process so much power all at once, defaulting into circuit overload unless things unfold at a more reasonable pace. But whatever caused this suspension of ordinary time, it was felt by everyone who encountered giant waves, especially the riders. Brett Lickle had described it as "like a car accident. A ten-second experience that takes two minutes in your mind."

When the lip landed with a whomping roar, the whitewater explosion alone was taller than many of the waves we'd seen today, a forty-foot geyser of aerated water. Because water is eight hundred times denser than air, a surfer trapped beneath a thick layer of whitewater—which is essentially foam—had only one hope of making it up for a breath: his flotation vest. This was precisely why the vests had saved so many lives. Without one a surfer could try to claw his way back to the surface, but it would be like clutching at mist. The superbuoyant vests changed the rules, popping up a two-hundred-pound human despite the lack of water density to support him. There, at least somewhat above water, he might get a gulp of air before the arrival of the next annihilating wall. It was also possible to suck a breath out of the foam, should it come to that. "If you're on the surface and you find air, baby, you're taking it," Lickle said, outlining the method:

"You keep your teeth closed and then make like you're sucking through a straw." In the aftermath of Killers' giant sneaker wave, several surfers had the opportunity to try this technique. As the men who had been caught inside began to surface, perilously close to the rocks, nearby Jet Skis shot in to help.

Everyone was so busy gaping at the wave that it took the captain a moment to notice that the surge had ripped the anchor off the boat, and we were drifting fast toward the impact zone. As he rushed to the wheel to right our position, the second wave in the set appeared, an only slightly smaller sibling. This wave had a rider on it: Coco Nogales. He'd been ready to tow and in the right position, so to everyone's delight the Mexican star got the star Mexican wave, and even if it was only the day's second biggest, it was still a pretty sweet consolation prize.

To meet Nogales and to hear his story was to root with all your heart for his success. A homeless runaway in Mexico City, he'd supported himself by selling gum on the street until, at age eight, he heard talk of a place called Puerto Escondido. It sounded idyllic, a small oceanside town south of Acapulco where kids could play in the surf, far from the urban zoo. He began to save up for a bus ticket. It took him seven months to earn the money. Upon arrival, he discovered that along with a less frightening existence, Puerto Escondido offered something else, something that, for him, would prove equally valuable in time: a wave that was a near replica of Hawaii's Pipeline.

From those beginnings, Nogales, now twenty-six, had scratched an unlikely path to professional big-wave riding, sponsored by Red Bull and others, and inspired a new generation of Mexican surfers. "Coco's really humble," Prickett said. "He lets his surfing speak for itself. I think he's gonna go a long way." Puerto Escondido, Prickett added, was such a hazardous wave, a ferocious, shallow beach break, that its riders had learned to be especially ocean savvy. "They're in bad situations all the time and have gotten used to handling them."

Now the waves were pumping and the tow surfers got their chance, but the rogue set was not to be repeated. Conditions stayed respectable, but the day's rhythm was erratic, a sudden blurp of energy followed by the doldrums, all of it shifting like quicksilver over the reef. There was general wariness; there were accidents. Brad Gerlach drove up on a Jet Ski and deposited Jamie Sterling at the boat, reeling from a burst eardrum. This injury, while painful and discombobulating—it destroys a person's equilibrium—was an ever-present hazard for surfers. It happened when a rider fell off a wave in a particular way, smacking down on the water ear first. The situation was potentially dire because when a surfer was trapped underwater with a cracked eardrum, he had no idea which way was up, and if someone wasn't right there to pull him out, he might not get out at all. On top of that the healing process was tedious, and the effects lingered. Spooked by the experience, some surfers took to covering their ears with duct tape whenever they were in the water. Sterling, his face pale, staggered onto the deck, accepted the offer of a Vicodin, and disappeared into the bunk area below.

By late afternoon the light was fading and so were the waves. Gathering our crew, we prepared to leave. Harro stood on the bow balancing a three-foot-long 600mm lens that looked like a piece of the Mars Rover. "It's all about being different," he said, showing me the unusual way the lens framed the wave. "Getting a take the other photographers don't have." He smiled, his face shiny with sunburn. Prickett climbed out of the water and set down his camera to take the icy Corona someone handed him. "So much chop," he said, shaking his head and drinking a gulp of his beer. "Man, did you see that one wave? It was getting scary out there." Jamie Mitchell, his wet suit replaced by a tracksuit, nodded. "All of the water energy was aimed toward the rocks," he said. "Like Jaws."

I turned and watched the wave as the boat steered toward Ensenada—and likely arrest for ignoring the harbor closure. No one seemed very concerned. The Mexican Coast Guard had surely seen its share of outlaw behavior, Prickett's floating package attesting to that, and anyway, big-wave surfers were troubled by legalities only before they'd gotten their rides. Afterward was another story: they were punch-drunk on adrenaline, blissed on endorphins, friends with everyone. Jail for a few nights? A four-figure fine? *Hagas lo que debes hacer, amigo:* Do what you gotta do, brah.

On the ride back I sat alone, staring absently at a pelican that had appointed itself our escort and replaying the monster wave in my mind: the defiance with which it kept rising and rising, far past the point where I expected it to stop, and the strange dreamlike way it moved, with a distinct aura of purpose. This too was part of the chase. During every big session there was always one wave that stunned everybody and would be remembered in years ahead. It represented what had been possible that day, and the man who had ridden it—if there was one—he would be remembered too. Every rider wanted that wave, even if he didn't say it quite so baldly, but he needed luck as much as skill to catch it. He needed to be in the precise spot at the exact time, ready to take off when The Wave appeared on the horizon. If he had just finished a ride, or he was rescuing his partner, or someone else's partner, or was in any of the wrong places he could be, doing any of the countless things he might be doing, that was just too damn bad.

When Hamilton caught his famous wave at Teahupoo, he'd been delayed getting out to the break because he stopped to help a friend who had misplaced his sunglasses. To fumble around on land for an extra twenty minutes on that day must have been excruciating, but if Hamilton hadn't done it, he might have been elsewhere in the rotation, not primed at the moment his singular wave raised its gorgon head.

We neared the marina and the pelican touched down, gliding onto the water. The wind had quieted again, and the sky was a rich cocktail of blues and golds. A spectral moon was on the rise, pale as smoke. Stepping onto the dock, I felt the woozy sensation you get when your feet hit solid ground after a day in the waves. Prickett, Harro, Mitchell, and Watson—up now for seventy-two hours and counting—were joking and laughing, replaying the highlight reel, as we made our way back to the trucks.

"Everyone's just like overgrown kids when it's big," Harro said with a grin. "This has been a whirlwind."

After a shower and a nap, there were more planes to catch and new swells to track. Prickett was headed back to Hawaii to work on a feature film called *The Warming*. The movie, he explained, was an eco-thriller about climate change. "The waters rise and rise and people die from big waves," Prickett said. "We're shooting dummy bodies getting trashed by surf on the rocks."

Harro was en route to Alaska, then Hawaii, and on to Australia, and those were only the stops he knew about. Tomorrow another megastorm could pop up and rip the tablecloth from under all of his plans. "Mother Nature rules all," Prickett pointed out. "You can't schedule anything, you just gotta wait for her to bring it." If she pointed you toward bandito-land, then that is where you went. Ensenada was a fairly unlovable location, a sprawling city of scruffy barrios, heavy traffic, gruesome time-shares, and bars that smelled of stale alcohol and armpits, but now it was also the place where I'd seen one of the ocean's grandest spectacles.

While they loaded the trucks, I checked my phone messages. Hamilton had called. He would be dismayed, I thought, by the accounts of this swell and what he had missed. His voice came on: *"Got your message, just getting back to you . . ."* There was a pause, as though he was grasping for words. *"Uh, we're still kinda recovering from our big day. Mentally, emotionally, physically. Spiritually. Intellectually."* Another pause while he took a long breath: *"Hopefully. Anyway, um, see you soon."* I listened, startled. Still recovering? From *what*? When and where and how was this a big day? Hamilton's voice had sounded drained, and the cadence of his speech was different, subdued. I stared at my phone. What on earth had happened in Maui?

I NEED THE SEA BECAUSE IT TEACHES ME.

Pablo Neruda

HEAVY WEATHER

SOUTHAMPTON, ENGLAND

Southampton, England, is a town that knows its ships. Located seventy-five miles southwest of London, it's a natural deepwater port that has been a maritime hub since the birth of Christ (at least). Ships of all kinds have been built, moored, exhibited, repaired, loaded and unloaded there; the port has hosted Viking raids and Roman conquests and French invasions and sorties for both world wars. It was from these docks that the *Mayflower* shoved off for Plymouth Rock and the *Titanic* sailed to its grim fate. For decades, the lordly ocean liners *Queen Elizabeth*, *Queen Mary*, and *Queen Elizabeth 2* began and ended their transatlantic journeys here. These days containerships come and go like clockwork, cruise ships are as regular as the tides, private yachts dot the marinas, and the town square is commemorated by an oversize anchor on a pedestal. When I visited Southampton, the place was beset by crowds who were there for its famous annual boat show, the showpiece of which was the tall ship from *Pirates of the Caribbean III*. But I was there on account of another vessel: the much-loved and well-used research ship, the RRS *Discovery*.

Along with its sibling ship, the *James Cook*, the 295-foot *Discovery* is berthed not in the port but directly in front of Southampton's National Oceanography Center (NOC), the two of them like a pair of very bulky cars parked in a custom-made driveway. NOC itself is a three-story facility built of gold-colored brick, and it stretches along the wharf, elegant and utilitarian at the same time. This curbside closeness to the sea— visible through every window and never more than a baby step away—is fitting: NOC is one of the world's most acclaimed marine research centers, home base for 520 scientists and staff, and 750 students from the University of Southampton.

One of those scientists is Penny Holliday, whose 2006 paper about the giant waves that bedeviled her research cruise aboard the *Discovery* had caught my attention. From the paper's provocative title ("Were extreme waves in the Rockall Trough the largest ever recorded?") to the bizarre incident it recounted (en route to Iceland, the *Discovery*, its crew, and twenty-five scientists had been trapped 175 miles off Scotland for a week in maniac seas), what I read was captivating and chilling in equal measure.

Holliday and her coauthors spelled out the science: how the ship had been ideally equipped to measure the waves, its shipborne wave recorder charting the ocean's every movement. They presented statistics for wind speeds, sea level pressures, and energy spectra. Charts and graphs plotted the wave heights, showing that the ship had run into several faces between ninety and one hundred feet tall. They put forth a theory as to why the seas had been so much crazier than the models predicted; it was due to an alignment of time, wind, and geography. The winds—strong, but typical for the region—had tracked the waves, traveling at the same speed and in exactly the same direction, relentlessly pumping energy into them across one thousand nautical miles. The result was a marauding, bulked-up gang of superwaves.

In her acknowledgments Holliday thanked the captain and crew for "enduring the terrible conditions" and "getting everyone home safely," and I had noticed this because few scientific papers end with a footnote stating that its authors are happy to be alive. As I read it, I was also interested in what the paper didn't say: how it felt to be caught in those waves and escape in one piece. Asking Holliday herself seemed the best way to understand the details of that ill-fated but revealing trip.

Penny Holliday handed me a cup of instant coffee and sat down at her desk, a scratched-up table in a bare-bones room. Reference books lined the shelves above it; a battered orange survival suit hung on the back of her door. At first glance, it was hard to imagine Holliday out on *Discovery*'s stern operating heavy equipment in fifty-knot winds. She was a tiny woman, strikingly pretty, with a sandy blond bob and ice blue eyes. Her laugh was light and bubbly. When she began to speak about her work, however, any inkling of fragility was gone.

Her specialty, the effects of climate change on ocean circulation, required her to spend extended stretches of time at sea, often at extreme latitudes. "Most of my research cruises have been in the stormy North Atlantic," she told me, describing how the currents that flowed into the Arctic were of particular interest. Both the temperature and the salinity of these waters have spiked upward in the past thirty years, since scientists first began taking measurements. With this flux comes the fear that the Gulf Stream will alter its behavior, scrambling the weather patterns in highly undesirable ways.

The burly, fifty-mile-wide Gulf Stream, which charges north from Florida before veering east at Newfoundland and making for Ireland, transfers heat from its eighty-degree waters to the North Atlantic, moderating the climate. It's part of a vast circulation system known as the Global Conveyor Belt, in which ocean currents loop around the planet powered by wind along with differences in temperature and water density, as they transfer the sun's energy from the equator to the poles. (In the North Atlantic alone, this process disperses a million power stations' worth of heat.) One of the white-knuckle questions about climate change is whether the Conveyor Belt will slow down—or even shut off completely—when the ratio of warm to cold water tips past a certain point.

Scientists have found evidence that this has happened before, as recently as the mid-1800s, and that for much of western Europe the off-switch resulted in a wild and icy ride. (Provence, France, for instance, would become as wintry as Maine.) All of which puts new urgency into Holliday's investigations.

Another woman who had earned her survival suit was Dr. Margaret Yelland, Holliday's office mate, seated across from us. Yelland's work called for an even greater dose of Dramamine, possibly an intravenous drip. She sought out the strongest winds available to perform her research on the ocean's ability to absorb CO_2, a critical function. "I've spent the last ten years of my career looking for high wind speeds," she said in her soft, husky Manchester accent. "Being sick as a dog in the North Atlantic and Southern Ocean winters. We're not really bothered where in the world we are as long as we get some good storms." Though Yelland hadn't been present on Holliday's infamous cruise, she'd placed some automated wind-measuring instruments on the ship and could thus monitor conditions from afar. "We were getting the weather data as it came in," she said. "I saw it and I just thought, '*Oh my God*.'"

Holliday laughed. "We kept thinking it was going to improve! The forecast kept telling us better weather was on the way." To say the least, this did not happen. From its start on January 28, 2000, the trip was plagued by an escalating series of storms. *"Several trashed cabins and one broken Mac monitor,"* Holliday e-mailed a colleague after her third night at sea. *"The worst Mac we had, but still, rather annoying."* Experiments were postponed while *Discovery*'s captain, Keith Avery, coped with conditions. No one had expected the winter North Atlantic to be a joyride, but neither did the ship's passengers realize they had signed up for the Cruise of the Damned.

It was a route that Holliday had traveled many times before, a transect called the Extended Ellett Line, after David Ellett, the Scottish scientist who began it in 1975. The Line stretched 1,200 kilometers from Rockall Island to Iceland; at stations along the way Holliday and the others would monitor the water's salinity, temperature, and composition. Scientists made the three-week trip annually, trying to understand the sublime balance of ocean and atmosphere, how things were shifting, and in general what was going on out there. "It's very disappointing how little we know about the ocean," Holliday said, echoing the sentiment of every marine scientist I'd spoken to.

Attempting to move from the Scottish shelf out to the deeper waters of the open ocean, *Discovery* made sporadic progress, the scientists getting a task or two done, only to be halted when the weather growled again. *"We have not been able to work for three days now and have experienced quite extreme conditions,"* Holliday wrote to a friend at the beginning of the second week. *"Because of that we have just remained hove to, which means steaming slowly into the wind and waves, trying to minimize the motion of the ship."*

Confined to their cabins and the lower deck areas, the scientists attempted to work at their computers, but this proved both futile and dangerous as furniture and other heavy objects catapulted about. "Chairs would fling themselves at you from unexpected places," Holliday recalled. "People were getting hurt. They were breaking ribs and getting covered in bruises and tossed around." Sleep was out of the question. For the most part, so was eating. Holliday struggled with a persistent low-grade nausea, not seasickness exactly, more like "having a permanent hangover despite not drinking much." But for many of the others, their discomforts exceeded the physical. "I think some people were struggling with the anxiety of actually being on the ship," Holliday said. "But if you start thinking, 'Oh my God, we're all going to die!' that's no way forward."

What came next wouldn't have helped them much: in the already wretched North Atlantic, the thirty- and forty-foot waves began to build dramatically, rising up to sixty, seventy, eighty feet and beyond. "It was absolutely awful," Holliday said. "We were being battered by waves that made the ship jump and shudder. The waves would loom up in front of the ship; we'd potter up them and seem to hover at the top before crashing down the other side. The scariest bit is when you are up there, looking into this enormous hole in the sea below. You imagine the ship might just continue down and not come up again. Sometimes we would drop suddenly and my eyes couldn't keep up with the motion. I lost my visual grip on the scene in front of me and my head was spinning."

Discovery emanated the creaks and groans of a haunted house, wood and metal stressed to the breaking point. Terrifying to be sure, but Holliday was also aware of a surreal magic to the scene, like being suspended inside an abstract painting made of salt water. Wind-whipped clouds of spray gusted from the wave crests, creating an aquatic whiteout, while seabirds spun eerily, like bats, overhead.

"There was violent movement on the bridge," Captain Avery recalled in an interview later with *Professional Mariner* magazine. Although other officers and engineers on the ship had argued for beating a hasty retreat toward shelter, any kind of relief, Avery knew their only hope was to point *Discovery*'s bow directly into the waves. Maintaining this position in such confused and angry seas was easier said than done. "It was particularly serious at night," Holliday said, "because the waves were not all coming from the same direction. So if you had a wave coming off from the left, you couldn't see it until it was practically upon you." For hours that stretched into two days, Avery battled with giants.

It couldn't have boosted morale when a lifeboat sprang loose in the night during a thirty-degree roll and began hammering *Discovery*'s starboard side. Or when a six-foot lab window suddenly shattered. "To me, that was evidence that the ship was twisting," Holliday said. "Which was very alarming because it had been lengthened in 1992. They'd cut it in half and welded in a new section. And so you're thinking about that . . . 'Hmmmm, this is the biggest test it's ever gonna get.'" She laughed and looked over at Yelland. "I've seen the movie *The Perfect Storm*. Liked it. But I never thought I would *live* it."

After nearly a week of maintaining the hove-to position, the seas subsided enough for *Discovery* to turn and dash for shelter. "We kind of surfed the waves back to Scotland and hid behind the Hebrides for a while," Holliday said. But volatile weather met them even there, dishing up a mix of hail and gale-force winds. When another major storm loomed on the weather maps, threatening to unleash a fresh platoon of thirty-foot waves, it made sense to end the cruise early, before the next beating started. To stay out any longer would be pushing *Discovery*'s already remarkable luck. *"It's all gone to rat shit,"* Holliday wrote bitterly to a colleague, lamenting the lost days at sea. Others were simply relieved that the scientists and crew had made it out of the waves without tragedy. *"Looking forward to seeing you all back safely,"* Holliday's boss, Raymond Pollard, e-mailed. *"That is the most important thing now."*

Disappointed not to have finished her survey of the Ellett Line, Holliday wasn't initially excited about the consolation prize: examining the data from *Discovery*'s wave recorder. "Lots of people said, 'Oh, you must write something about those waves,'" Holliday said. "But I never really got round to it. And I'm not an expert on waves." It was only in 2005, when another paper was published touting the ninety-one-foot waves that were measured in Hurricane Ivan, that Holliday's competitive spirits were piqued: "I thought, 'Hmmm, we've got bigger waves than that.'"

Yelland urged her on, interpreting the wind measurements and crunching the numbers. Even with countless hellacious ship days under her belt, she was floored by what Holliday and the others had endured.

"And we weren't even in the windiest spot!" Holliday said, pointing to a graph on her paper that plotted wave heights and wind speeds. "So we probably weren't in the place where the waves were at their highest." She flinched behind her desk. "But they were high enough for me!"

Though the relationship between wind and waves has been well charted by science, neat formulas demonstrating that if the wind does *x* then the ocean will do *y*, one of the most intriguing aspects of *Discovery*'s ordeal was that the biggest waves did not accompany the strongest winds. Rather, the hundred-footers showed up more than a day after the most violent gusts had subsided, at a time when the scientists believed the worst was over. "The point is that all of these previously measured [giant] waves were under hurricane conditions, really extreme conditions," Holliday said. "But our big waves weren't."

All of which begs the obvious question: What, then, *did* cause the "largest ever recorded" extreme waves?

Holliday and Yelland believed it was an effect known as "resonance," an aspect of nonlinearity that is endlessly complex when scrawled across a whiteboard and kindergarten-simple when explained by the analogy of a kid pumping his legs on a swingset, dramatically boosting his height on each pass. Energy is continually being added to the system, more and more and more, in erratic bursts, until the swing can go no higher. Likewise in the North Atlantic, wind energy surged into the waves until they grew to enormous proportions. "The wind was actively forcing the growth of the waves for a very long time," Yelland said. "So they kept building and building and building."

"The wave model that we were looking at didn't predict the waves that we actually encountered," Holliday added. "It got the wind speed right, and it got the arrival time of the waves right, but they were much smaller than the actual waves we measured. So the implication—the *concern*—is that these big waves are out there, and if the models aren't reproducing them, then the engineers who are using the models to design their ships or whatever, well, they might not be looking at the right limits."

She paused to let this statement sink in. "I went to the library and looked through fifty years of synoptic weather charts," she added. "The conditions we encountered were not that unusual. And so I think what we are trying to say is that these waves happen more often than we realize. We're just not measuring them."

During my time in Southampton I often felt as though I'd dropped into a parallel universe concerned entirely with waves and water; a place where everyone spent their days thinking about the ocean and yearned to uncover its most closely guarded secrets. In every corner of NOC, behind every stack of books, tucked into a warren of offices and conference rooms and libraries and labs, there was someone devoting his or her life to grappling with the mercurial sea. Rapid-climate-change specialists and tsunami experts crowded into the quayside cafeteria with submersible operators and marine biologists; wave physicists and sonar technicians nodded to storm modelers and ship captains in the halls. Brilliantly colored bathyspheric maps—depicting the slashes, trenches, and subduction zones rent into the seafloor, the lively underwater places where the earth is shifting and splitting and bashing into itself—lined the walls. "We know more about the surface of the moon than we do about the deep ocean," the center's press materials read. Its daily activities aimed to change that state of affairs.

"I tend to work on landslides and geohazards of the deep sea, from which tsunamis are spawned," Dr. Russell Wynn told me by way of introduction. Though I had come to Southampton to learn about Holliday's voyage, I also wanted to meet as many of the resident wave experts as I could. Wynn was a tall, lanky guy in his mid-thirties with sleek features and an intense presence. He sat in a spacious office with graciously high ceilings on the institution's top floor. These days, he explained, tsunami science was hopping due to technological advances and increased interest after the horrific Indonesian waves of 2004. Suddenly funding had become available to determine threat levels elsewhere. "We're coring all the way along the European Atlantic margin to Northwest Africa," Wynn said, describing the process of drilling a probe into the seafloor and then examining the earth's layers to discover what furious geological events had previously occurred there. From those findings it would be possible to deduce the odds of similar upheavals in the future.

This information matters more to western Europe than most of its inhabitants know. Though the Pacific is credited with all manner of killer-wave potential, the Atlantic has hosted its share of trouble. Prehistoric evidence depicts much giant-wave damage along the coast of Scotland. Sixty-seven tsunamis have struck Italy in the past two thousand years; in 1908 a powerful one in Messina killed eight thousand people. Devastating tsunamis have also hit such unlikely spots as the Virgin Islands (1867); Nova Scotia (1929); and even Monaco and Nice along the French Riviera (1979). In 1755 a tsunami leveled Lisbon, Portugal, killing sixty thousand people. The 8.8 earthquake that caused it was felt all the way to England; at its epicenter the waves it produced were more than fifty feet high. The tsunami had also surged north, busting into Irish harbors and British ports, and east, causing death and damage as far away as Puerto Rico.

Wynn had also logged much time studying past volcanic collapses in the Canary Islands, and he disagreed with Bill McGuire's notion that the Cumbre Vieja volcano threatened the entire Atlantic basin with a mega-tsunami: "We've sort of locked horns a few times." After having identified, cored, mapped, dated, and analyzed the seabed deposits around La Palma, Wynn believed the west flank of the island would crumble in bits and pieces, resulting in far smaller waves: "What the mud tells us is that these things are not as big a hazard as has been proposed." He clicked on his laptop and pulled up a 3-D animation of La Palma's western flank falling into the sea, rendered in pretty shades of green. It looked as though a set of giant jaws had bitten off half the island, spitting bungalow-size blocks of rock onto the ocean floor. Even at diminished size the resulting waves would make for dark days in the Canary Islands, but they wouldn't make it across the Atlantic or even to British shores. But while Wynn did not buy into the severity of this particular scenario, he did second another one of McGuire's concerns: that climate change will lead to increased tsunami risk worldwide.

"The sea level going up and down has a big impact on landslides," he said, sitting at his desk, his hands clasped above a sheaf of papers and a book titled *Surviving the Volcano*. "And it can also have a big impact on the number of earthquakes you get." Wynn's voice was crisp and authoritative, and he spoke with the measured calm of someone who dealt in geologic time. There might be disasters, yes, but they would be handled. The content of what he was saying, however, left a different impression.

Geologists now knew, he told me, that when the last ice age ended about ten thousand years ago it had not gone quietly, but rather in a flurry of seismic fits. Millions of tons of melting ice drastically hiked sea levels, causing the entire ecosystem to go tilt. The ground trembled and shook; volcanoes that had been sleeping for aeons suddenly sparked to life. As the planet's chemistry and equilibrium were knocked askew, the oceans seethed. It was not hard to see the parallels between that time of upheaval and our current situation, in which glaciers are dwindling at a startling clip. "Isostatic rebound," McGuire had called it, a simple principle with awful implications.

"You start to load more water onto bits of the seabed that might not like being loaded," Wynn explained, "and therefore they fail—that failure being in the form of an earthquake. I mean it doesn't sound like much, but if you raise the sea level by a centimeter, and you extrapolate that centimeter of water across several hundred thousand square kilometers of sea bed, that is actually an enormous, enormous load. And at the moment, well, the sea level's starting to go up quite quickly."

"Quickly?" I asked. I'd heard scientists describe the oceans as rising "steadily," and "inevitably," but no one had put it in quite such immediate terms before.

"Yes," Wynn said. "And it's only likely to accelerate. So it could be that we may enter a phase of more instability of the seabed."

If you were looking for the perfect set of circumstances to create memorable tsunamis, a fast-changing, unstable, undersea environment would be at the top of your checklist. The relationship is straightforward:

when large batches of rock and sediment move around down there, mayhem is uncorked on the surface. Along with any extra earthquakes we can expect to roil the bottom, there is also another concern. "The land is eroding much faster than it did, say, a thousand years ago," Wynn noted. "There's more sediment being shed into the ocean now." That silt, sand, soil, and other material then heap up underwater, a stockpile of extra ammunition for the next slide. "So the ocean's stormier and the sea level starts to rise; there's increased load on bits of the seabed and bits of the earth, and the possibility for future landslides is greater." He sighed. "It doesn't take a genius to work it out. Sooner or later these things are going to come back and bite us."

For a demonstration of just how destructive landslide-induced tsunamis can be, scientists point to the Storegga Slide, a catastrophic slump in the North Atlantic that happened about 8,200 years ago. A section of Norway's continental shelf the size of Kentucky gave way, plummeting down to the abyssal plain and creating a series of titanic waves that roared forth with a vengeance, swiping all signs of life from coastal Norway clear to Greenland (and reaching as far south as England). The Shetland Islands were especially hard-hit, by tsunami waves that likely measured over sixty-five feet high. Farther south, the waves drowned a Wales-size landmass that connected Britain to the Netherlands, Denmark, and Germany. (In other words, Britain wasn't always an island.)

So what caused the underwater slide that produced the Storegga waves? Scientists aren't entirely sure. An earthquake in the North Atlantic, perhaps. But there's another, eerier possibility. It may have been due to a blowout of methane hydrates, gas deposits that are frozen into the seafloor. These gobbets of ice, which look like tiny snowballs (but generate a gas flame when lit), carpet the world's oceans. In particular they are clustered on the continental slopes, ideal for landslides. Methane hydrates are hair-trigger sensitive to changes in pressure and temperature: one extra degree is enough to melt them. When they release, they not only collapse the seafloor around them, causing landslides, they can burp vast clouds of methane into the atmosphere—a greenhouse gas ten times more potent than carbon dioxide. As for how much methane is currently (and, for now, safely) frozen down there, the United States Geological Survey conservatively estimates that these submarine ice balls contain twice the carbon to be found in all known fossil fuels on earth.

It was Wynn's job to spin present-day fact from these geological mysteries of yore, to apply rigorous science to the "what-if" disaster scenarios and pull solid probabilities out of fearful conjecture. "We're the guys who say, well, in this particular area tsunamis are going to happen every hundred years, or whatever," he explained. In the Canaries, for example, Wynn and his colleagues figured that major slides and tsunami events took place approximately every 100,000 years. (The last one took place 15,000 years ago.) There was a disclaimer on that return rate, however: "For geologists nothing's certain. So I couldn't put my hand on my heart and say that a big chunk of the Canaries will not fall into the sea tomorrow. You just don't know."

Another nearby area that had captured scientists' attention was Spain's southwest coast. "There's quite a bit of seismicity around there," Wynn said, "and it's a heavily populated region." After all, the 1755 quake that "clobbered Lisbon hard" had been a superheavyweight. The shaking—which lasted for almost ten minutes—spawned fifty- and sixty-foot tsunamis that wreaked havoc from Morocco to England. "Can we expect another one of those in the next ten years?" Wynn asked rhetorically. "Hundred years? And which parts of Europe have been affected by similar events in the past?" These questions were deemed urgent enough that Wynn was about to embark on a monthlong research cruise, one of several planned for the near future.

There must be something etched into human DNA that allows us to quickly forget such shattering events as the Lisbon tsunami, which upended the lives of millions across all of western Europe and northern Africa—"two hundred and fifty years ago," Wynn pointed out, "that's nothing." Then there was Krakatoa's supereruption and tsunami in 1883, so recent that its 140-foot waves coincided with the premier issue of *Ladies' Home Journal*. But while our collective memory may be woefully short, the geological record is long. "By studying deep-sea deposits we can start looking at the past landslides in the area," Wynn said, sweeping his hand across a map of the Atlantic margin. "We can tease out the earthquake record as well. That's what we're doing out there. We're trying to unravel the history of these waves."

"So where do you want to start?" Dr. Peter Challenor asked. "We've done quite a lot on extreme sea states. How much do you know about wave statistics?"

Challenor spoke quickly, punctuating his speech with sharp birdlike gestures. He seemed to take genuine pleasure in discussing his work. Even the greenish fluorescent lighting in his office couldn't dampen his exuberant aura. His brown hair grew lavishly onto his face, happy curlicues of sideburn and mustache and beard. Across from him sat his colleague Dr. Christine Gommenginger, in a smart navy blue dress. Behind him, a blank whiteboard beckoned.

The two scientists specialized in remote sensing of the ocean, gleaning snapshots of its behavior from space. In particular, they examined the waves. Eight hundred miles up in the exosphere, the European Space Agency satellite known as Envisat zips around the planet fourteen times a day, shooting radar pulses down onto the sea surface. Using the information it (and other satellites) send back, Challenor and Gommenginger can chart wave heights anywhere in the world with ridiculous precision.

It wasn't always this way. Before 1985, when a satellite called Geosat was launched, wave scientists had to rely on moored buoys and reports from ships for their data. Better than nothing, perhaps, but given that the buoys were clustered near coastlines and the ships could survey only pinches of ocean real estate, what was going on out there was really anybody's guess. The latest in a series of increasingly sophisticated satellites,

Envisat is the largest earth-observation spacecraft ever built. Resembling something dreamed up for a James Cameron movie, packed full of powerful instruments with impressive acronyms like GOMOS (Global Ozone Monitoring by Occultation of Stars), DORIS (Doppler Orbitography and Radiopositioning Integrated by Satellite), and ASAR (Advanced Synthetic Aperture Radar), there's not much it can't do or see. Thickness of the Arctic sea ice? Surface temperature in the Somali Current? Size of the waves off Peru? No problem.

"We tend to concentrate on significant wave height," Challenor told me. Rather than pinpointing individual waves, this number, the average height of the top 33 percent of the waves, paints an overall picture of surface roughness. Over time scientists can use it to determine a key statistic for any given patch of ocean: the size of the "hundred-year wave." Theoretically, only one wave larger than this value (on average) should show up every century. "We produce this mainly for people who are building structures and want them to survive big waves," Challenor said. That included oil rigs, of course, along with coastal construction and a relatively new concern: wave farms. "Like wind farms," Gommenginger explained. "Wave energy is coming."

This low-impact form of alternative energy seems brilliant on paper, but in the past wave farms have not fared well. The devices meant to float at sea and capture the waves' power have been destroyed in short order by . . . the waves. "They've all been smashed up in storms," Challenor said, shaking his head. "That's usually their fate around the second or third winter." Heartier designs looked promising, but finding the ideal location was also part of the deal. "They want it rough," Gommenginger said, "but not too rough."

According to Challenor the waviest places of all were "the North Atlantic in winter. Or the Southern Ocean anytime." There ships could expect to encounter thirty-foot seas on a good day. That was not to say, however, that monster waves couldn't make surprise appearances in other locales, at other times. (And by the way, building to withstand the hundred-year wave wouldn't help anyone who happened to meet up with the thousand-year wave.) "The way the radar system works, the very big ones are difficult to measure," he said. When behemoth waves appeared in the satellite data, the space agencies considered these readings to be errors, and they were automatically deleted. "They give you missing value code instead, which is really annoying. We shout at them for that."

In general, I wondered, did they agree with Penny Holliday that these exceptional waves—in the ninety-foot range and beyond—were more common than people realized?

"Okay," Challenor said in a brisk voice, as if we were finally getting down to business. He hunched over in his chair, nodded soberly, and crossed his arms. "That's a good question. *Yes,* I do. There are a lot of high sea states. You don't hear about them, because people don't go out into them. Very sensibly." Adding to the waves' elusiveness, the instruments deployed at sea to measure the giants were usually pulverized in the line of duty. Oil platforms took some heavy abuse that could yield clues to maximum wave heights, but the oil companies tended not to report it. Challenor, who began his career as a wave statistician working on North Sea rigs, had experienced this firsthand. "I would ask them if there had been any damage, you know, trying to get some feel for it. And they'd never tell you. But if you looked in the industry papers, and at the insurance claims . . . things were getting smashed off." He chuckled at the memory. "There was an awful lot of commercial secrecy back then. That's changing."

For wave scientists, these were bullish times. "In the last few years studying waves has suddenly become fashionable," Challenor said, looking bemused. "Used to be we were considered quite odd."

"Yes," Gommenginger agreed. "Until two years ago we were laughed at because we did waves." The ability to measure the ocean from space had attracted interest, she thought, and that new information had increased demand for improved climate models and forecasts. "It's a combination of factors all coming together."

"And this whole climate change thing," Challenor added in a serious tone. "My betting is that the waves will get worse. But it will vary from year to year. It's not a simple relationship." An atmospheric pattern known as the North Atlantic Oscillation (NAO) was predicted to rise, he noted, heralding stormier days ahead: "The waves are driven by this." (From 1963 to 1993 a strong NAO increased wave heights in that ocean by 25 percent.) Going forward, if the seas responded the way many scientists feared they would, Challenor said, "we do have to do something about ships and oil rigs. Topping off dikes. Coastal erosions."

"There are still fundamental things we don't understand," he added, pointing out that a certain kind of freak wave—a mutant that was three or four times taller than its surrounding seas, without any obvious cause—still could not be explained by science: "We don't have the math." It was all very well to make freak waves pop out of a laboratory tank, "but what happens in the real world where everything is random and messy?" He stood and turned to the whiteboard. With a red marker he sketched a diagram of a giant wave spiking out of a much smaller group. "We don't have that random messy theory for nonlinear waves. At *all*. We don't even have the start of a theory!" He put down the marker. "People have been working very actively on this for the past fifty years at least."

"What is it we're not understanding?" I asked.

"We don't know!" Gommenginger said. The two scientists laughed.

"My suspicion is the nonlinear part," Challenor said, his voice revving faster. "Which is what the real waves out there are! You don't just get one wave interacting with another wave, which would be hard enough. There are three-way interactions—and I suspect there are four- and five-way interactions. Bits of wind, odd waves coming in from other directions, things bumping into it . . . So you've got this whole messy random field that's just talking to each other in this horrible way that we don't understand."

"We might need a different tack," Gommenginger said. "Someone from a completely different discipline to come."

"Maybe some other area of physics," Challenor agreed.

Gommenginger smiled. "But as we said, until a few years ago there wasn't much interest in waves at all. We're just coming up." She looked around at the room, a colonnade of papers, books, and computer printouts stacked on every available surface. "Watch this space."

Andy Louch sat in his office on NOC's ground floor, overlooking the research ships and the commercial port beyond. Hefty binoculars stood propped against the window, alongside a satellite phone. Maps and boat pictures lined the walls. Louch was solidly built, with fine chestnut hair and a broad, friendly face. A veteran seaman for twenty-seven years, the former captain of the *Discovery*, and now the center's operations manager, he knew the perilous shipping landscape inside and out. Louch logged countless hours in the North Atlantic in his day; he had traversed the Southern Ocean off Antarctica, thousands of miles from land, with only "growlies," piano-size slabs of ice, to look at while the lunatic winds and the bunker swells rolled by. He understood the wave misbehavior Avery and Holliday had faced; he too had witnessed the angriest seas with their hundred-foot soldiers, the kind of immemorial fury that inspired author (and former ship officer) Joseph Conrad to write, "If you would know the age of the earth, look upon the sea in a storm."

Even so, Louch lacked the slightest hint of swagger. His descriptions were factual, understated. His voice was so calm that listening to him describe the *Discovery*'s ordeal was almost a soothing experience. "That cruise was exceptional," he said. "It really was. They just had so much bad weather."

Getting caught out in a hoary North Atlantic storm is nobody's idea of fun, but Louch maintained that with proper safety procedures—as demonstrated by Captain Avery—ships could hold their own. The trick was not to get impatient, not to budge from that hove-to position. "We've always been very pragmatic about that," he said. "You just sit it out. It's too rough to do any science anyway. The ship's moving around quite violently, it's fairly uncomfortable, but you're relatively safe." He gave a half shrug and grinned. "And yeah, okay, it can be a bit worrying at times. You see the big waves coming toward the ship and everything else."

In general, Louch emphasized, the ships that ran into problems were those that decided to press on during a storm. One advantage of a research cruise was that it wasn't on a tight timetable, where millions of dollars would be lost if the ship was delayed. "For a captain on a commercial ship, the pressure is enormous," he said. "Obviously, you want to be in Quebec, or wherever, on whatever day it is you've got to be there. But you also don't want to burn too much fuel, and you don't want to damage the ship or, primarily, the cargo." He gestured toward the port. "Whatever you do, you don't want to lose your batch of expensive cars."

"The cars *fall off?*" I'd heard stories about thousands of tennis sneakers and rubber toys bobbing around out there after toppling from ships, but somehow I hadn't imagined a mass graveyard of Porsches.

"Oh, absolutely. It's not uncommon."

Despite the obvious terrors of hundred-foot waves, Louch maintained that some of the fiercest conditions involved stretches of shorter, choppier waves, in places like the Baltic Sea. "You might only have five- to ten-meter seas but the period is every ten seconds, so it's constantly battering, really quite vicious," he said. "It's easy to underestimate them."

In every corner of every ocean there were hazards that demanded more precaution, an extra dollop of respect. In the Southern Ocean, it was the isolation—"If you do have a problem you've got no one to call on"—along with floating ice, an ever-present concern. "The rule of thumb for ice, basically, is that bergs the size of a house and upward are not a problem," Louch said. "You can pick those up on radar. The smaller bits are more dangerous. It's like hitting solid rock." In the North Atlantic it was the potential for extreme wave heights: "If you get a bad wave it can actually roll you over. Turn you turtle." Any captain who made the mistake of pointing his stern into a big swell would find little mercy.

"The bottom line is that a ship is a big steel box," Louch said. "As long as you keep the air inside it—you keep your hatches secured and battened down and your doors shut, you're okay, generally. It's when the weather is so severe that you actually break things—knock the hatch covers off or get fractures in steelwork or have structural failure . . ." He left his sentence unfinished.

There was one storm from decades past that seemed to haunt Louch, "a big, midwinter Atlantic depression" that had threatened a research ship he was commanding, the *Shackleton*. "We sailed out of Gibraltar," he said, "to do some work on the mid-Atlantic ridge. Put some oceanographic moorings down." Even then, back in 1978, the weather radar was able to pick up a storm of this size. "But sometimes you can't get out of the way. It may be four, five hundred miles across." Like the *Discovery*, there was only one thing the *Shackleton* could do: brace itself. "We had three, four days when we were stuck out there," Louch said. "But there was another ship sank, only two hundred miles from us." A stricken look came over his face. "A big containership. She was ten times bigger than we were. We never knew what happened to her, but I assume she was trying to push along, probably doing seven or eight knots, and really crashing into the waves." He shook his head sadly. "They lost her with all hands. We never heard any distress signal. We couldn't get near to where her last position was. The weather was too bad for us to even head two hundred miles to get to her."

It was the kind of story that didn't go away, especially if you'd been in those shoes yourself, scared in the dark, powerless in the waves, overwhelmed by the screaming wind and the ceaseless pitching and rolling. "What was the ship's name?" I asked, almost as an afterthought.

"She was the *München*."

I stared at him, remembering the nightmarish things I'd read about that vessel's demise. The desperate search. The empty lifeboat, drifting. The torn and twisted metal that spoke of forces beyond comprehension. *"Something extraordinary"* had destroyed the ship: that verdict had haunted me too. With Louch's words freshly stamped on my imagination, I found it unsettling to leave the building and see the crowded docks, the autumn-colored city of containers, ocher and rust and dull blood red. The containers bore the names of faraway ports as well as the companies that plied them—Maersk and Hyundai and Hapag Lloyd—the modern world sharing space with the ghosts of the *München*, the *Titanic*, and the countless other ships that left from this place, turned into the waves, and never came back.

Tow-in surfers Brett Lickle and Laird Hamilton were catching some of the biggest waves ever ridden Monday . . . History was in the making. And that's when things went wrong.

Honolulu Star-Bulletin, *December 6, 2007*

EGYPT

HAIKU, MAUI

It was completely unforecast," Hamilton said, taking a seat at the picnic table outside his garage. He set down two cups of espresso and slid one over to me, a little midafternoon jolt. "I went and looked at Pe'ahi first thing—and I could tell that the angle was shitty. Jaws doesn't like a north swell. That's when we made the call to go to Sprecks." He looked at Brett Lickle, sitting across from him. "Egypt," Lickle said, remembering. Lickle took off his baseball cap and pushed his hair back. His pale blue eyes, I noticed, had taken on a wary, haunted look. From what I was hearing, there was very good reason for that.

As it happened, the early December storm hadn't bypassed this island at all. In fact, though the storm's waves had been impressive at Mavericks and Ghost Tree and Todos Santos, the true measure of its power hadn't been felt on the far side of the Pacific—it had been unleashed eight miles from where we were sitting. On December 3, not only did the swell rampage over Maui, it hit here hardest of all. When it did, Hamilton and Lickle had been squarely in its crosshairs. I could get only the basics of the story out of Hamilton by phone, so I'd returned to Maui to fill in the details. That day he and Lickle had experienced the closest call of their careers.

"It wasn't real big in the morning," Hamilton recalled. "And I don't remember it as being very stormy. It was nice, actually."

"Nobody had any idea," Lickle said.

"It got more and more ominous as the day went on," Hamilton continued, "as the front came in with the surf. As the surf grew it became dark. The clouds got thicker and denser. And denser. And darker."

At nine-thirty that morning the pair had launched from Ilima Kalama's place (Dave Kalama's father) on Baldwin Beach. Dave himself was out with a torn calf muscle, hobbling around in a leg brace. Nothing dramatic was expected, so no photographers had been called to the scene. It was simply an average tow day, better than nothing. As Hamilton and Lickle motored out toward Spreckelsville, they encountered only one other tow team, their friends Sierra Emory and John Denny.

Emory, a world-cup windsurfer and big-wave prodigy, lived next door to Hamilton. He too had been on the scene at Jaws during the early experimental days. Denny was a local, and he often towed with Lickle. The four men were surprised to find themselves alone on what looked like a solid forty-foot day, but they chalked it up to the fact that nothing had been broadcast; there had been no alerts on Surfline or any of the other forecasting sites. Also, the swell's unusual direction favored those who knew the idiosyncrasies of Maui's outer reefs.

"And at that point you had no clue?" I said.

"Did I know what was coming in the afternoon?" Hamilton said, shaking his head. "No way. Had I known I would have gone straight home and rested to get ready. I wouldn't have been out using energy."

For the first two hours he and Lickle traded off driving and surfing. Although he'd felt sick in the morning and was less than enthusiastic about being out there, Lickle rode a few waves and began to perk up. "Then all of a sudden it started building," Hamilton said.

"It just kept coming up," Lickle added, in an emphatic tone. "It got bigger and bigger and *bigger*."

The waves were growing so radically that they sped back to shore to change their equipment. "I needed a different board," Hamilton said. "I wanted my gun." He paused, and when he spoke again, his voice had lowered an octave: "Then when we got back out there, we were like, 'Ohhhh, shit.'"

The phone rang, and with that Hamilton tipped back the last of his espresso and went into the garage to answer it. Lickle and I stayed outside. A few minutes later we were joined by Teddy Casil, who drove up on the Mule, one of Hamilton's beat-up off-road Jeeps. Speedy and Buster leaped from the backseat and ran to us with the joyous, pogo-stick energy that only dogs possess. Herded up the hill by Casil, Ginger and Marianne, all six hundred pounds of them, ambled over the ridge and plunked themselves down in a mound of red mud where they promptly fell asleep, side by side.

It was a gusty, sunny Wednesday, the wind having blown the surf to chop. When there were no waves there was a need to talk about waves, and as Hamilton continued working in the garage and the afternoon dwindled and nature's alchemy turned the light over the pineapple fields to gold, the picnic table filled up. Emory, sweat-encrusted from a day of landscaping, came over from next door. He was a friendly, scruffily handsome guy, dark haired and brown eyed, with the laid-back calm of someone who'd spent his entire life in Hawaii.

On Emory's heels, Don Shearer pulled up in his four-ton truck. Shearer was a helicopter pilot of such renown that a six-part TV series and a BBC documentary had been made about his exploits. For twenty-five years and counting he had rescued stranded hikers from the Hawaiian backcountry wilds, plucked people out of flash floods and riptides and gulleys. He'd recovered more than eighty bodies, victims of drownings, suicides, shark attacks, plane crashes, boat accidents, all manner of random mishaps and tragedies. From the cockpit of his canary yellow MD-500 helicopter (which, terrifyingly to a civilian, has no doors), Shearer had eradicated major swaths of Hawaii's marijuana crop and dumped buckets of water on raging fires. On the outsize days at Jaws he always flew safety, hovering low over the water to give the cameramen an ideal vantage point and poised to lower a medevac sling, should it come to that (and it usually did).

Shearer sat down and loosened his heavy work boots, reaching to take an icy Coors Light from Casil. Powerfully built with a shaved head and a no-bullshit commando look in his eyes, Shearer could be intimidating—until he smiled or talked sweetly about how crazy he was for his wife, Donna, or called you bruddah, which he eventually did to everyone he liked.

I had barely started in with my questions about December 3 but I knew that the full story would come to light only over time. Lickle was clearly still shell-shocked, and Hamilton was always monosyllabic when it came to his own feats. But I was heartened by Emory's appearance at the picnic table, knowing that on December 3 he'd been out at Egypt too. Just as I was about to steer the conversation back to that day, Shearer did it for me: "Let's see it, bruddah," he said to Lickle. "How's the wound?"

Lickle leaned over and hoisted his left leg to reveal an angry red railroad track that zippered down the back, clear from his knee to his heel. The scar was an inch thick, still swollen, peppered with divots from where fifty-eight medical staples had gone in and then—two days later—come out again due to staph and strep infections that had threatened to cost Lickle his entire limb. He'd then spent a week in the hospital, pumped full of antibiotics with his leg flayed open so it could be properly cleaned out before being stapled up again. As far as gnarly scars went, this one was a Hall of Famer. But when I looked at it, I didn't think of how bad it must have hurt or what a drag it was or feel sorry for Lickle. All I could think about was how lucky he was to have survived the wave that caused it.

By the time Hamilton retrieved his big-wave gun and he and Lickle made it back to Egypt for their second session, conditions had amplified drastically. It took them thirty minutes to drive the Jet Ski a mile and a half, jockeying with forty-foot waves breaking across the inner reefs in a solid wall. A boom of dark clouds had lowered and the air was saturated with salt and drizzle. Hamilton described it as the "worst visibility I've ever seen on Maui."

"Ah, cuz," Shearer said when I mentioned this. "The weather was just shit. Totally ridiculous. It was like, 'Where'd *this* come from?' Visibility less than a mile, clouds lower than three hundred feet. The airport was IFR—only one aircraft in the airspace at a time because they have to land on instruments. Hardly ever happens."

"It was spooky," Emory agreed. "Because it was overcast and dreary and gray and we didn't think it was that big. We rolled up to the beach thinking, 'Ah, it's nothing.' We got out there and it was like, 'Woah! Wake-up call! Holy shit!'"

Knowing the surf was rising, Shearer had picked up the injured Kalama, and the two headed out in the helicopter to fly safety for Hamilton and Lickle. First, though, they took a quick pass over Jaws. "It was all disorganized and funky," Shearer recalled. "I'd never seen it like that. It was really screwed up." They buzzed back over Egypt but in the veiled light they didn't spot the surfers—and they couldn't hang around to look because they were sitting right in the airport's landing path. "We could see big waves on the horizon north of Spreckelsville," Shearer said. "And we could tell there was something going on that we'd never ever seen before. I kept asking the tower for clearance to stay out there—they're all my buddies—but they couldn't approve it." Reluctantly, he and Kalama turned back to land.

Casil reached into the cooler and passed out another round. "That whole day just sounds cartoonish," he said.

"Oh, it *was*," Lickle said, popping open a can. "The bigger it got, there was no reality to it. It was the friggin' *Twilight Zone*."

From somewhere back in the garage, Hamilton chimed in: "It was another scale! Other scale. *Metric* scale."

When Hamilton and Lickle made it back out to Egypt, they were floored by what greeted them: snarling faces as large as any they'd ever seen at Jaws. And improbably, a left break farther out—a wave they had never known existed—was exploding at even greater sizes. Egypt, Hamilton said, was "taller than tall," with an oval lip that flared at the top like the hood of a cobra, extending the face. "It has a big pyramid shape," he'd explained. "It's really steep at the top and then it's just bottomless."

Amazingly, on December 3 there was little wind, leaving the water chop-free, with an oily, glassy surface. "They were the most perfect big waves I'd ever seen," Lickle said, shaking his head. "What'd you call it?" he yelled to Hamilton. "Butterball turkey butts?"

"Butterball turkey butts!" Hamilton yelled back, above the clanking of tools.

"Something went wrong with my board for the second session," Lickle said, adding that he'd picked it up to launch and noticed the fins had somehow gotten crinkled and bent. "I'm like, 'What the fuck?' And Laird goes, 'Here, take mine.'" Lickle laughed wryly. "I'd ridden his board once before so I had the confidence to jump on it. But it was a whole other animal: my board's a Corvette, but Laird's board—it's a Ferrari. When you kick out of a wave you're going fifty."

Given that Lickle was a good five inches shorter than Hamilton, the foot straps were a stretch. But he went for it—"I can't tell you how scared I was"—and rode what, even now, he referred to as "the wave of my life."

"When I kicked out of it," Lickle continued in an incredulous voice, "I said to myself, 'You're out of your fucking mind, buddy, you're *cut*." He made a curt slicing motion with his hand. "I remember just looking at Laird and going, 'Done.'"

"You scared yourself out of the water," Shearer said with a nod.

"That wave scared me out of the water," Lickle agreed. "I barely made it. I mean it was by the friggin' hair of my ass. But I was ready to watch Laird ride some of these bombs. 'Cause you don't get to see it while you're doing it."

Meanwhile, right before Lickle's ride, unseen in the mist, Emory had taken a fall at precisely the wrong moment—and most definitely on the wrong wave. "I was in exactly the worst spot," he said. "Dead center." A spooked look passed over his face as the memory resurfaced. "It was the biggest wave I've had on my head. A heavy, heavy one. I got on the sled and I'm seeing spots and stars. And it's dark and John's driving in circles looking for my board and I'm going, 'Which way is the ocean? Where's land?' "

"How big was it?" I asked.

"You know, in the moment I don't pay attention to the size of the wave," Emory said, deflecting the question. "Some are scary big, and some are just big." He looked at Lickle, his eyes wide. "Brett, my board—as fast as it is—I was going *backward*. The wave just got bigger and bigger and all I could do was keep going straight. It was as good as it could be, dead smooth, nuts clean, and I *still* couldn't go."

Lickle nodded. "And whole sections have gotten blipped out of your memory because things were so intense. You really don't know how big that wave was." He raised his eyebrows. "But that's good. You don't want to see too much, or you'll never get on the rope."

"I haven't had a schooling like that in a long time," Emory said, flinching.

"Well," Lickle said, turning to me, Shearer, and Casil to underscore what he was about to say, "his wave was a monstrosity. I remember looking at him on it. Laird and I were towing back out. Sierra was in the middle, and I could have stacked ten people above him and ten people below. Little stick figures in my mind."

At that point none of the four surfers could have imagined that Egypt was only getting started, that however harrowing Emory's spill had been, it was a warm-up act for what was yet to come. While Emory and Denny regrouped, Lickle had turned to the matter at hand: putting Hamilton on the craziest wave he could find. Even in the poor visibility he could make out the darker depressions as they approached, shadows etched into the ocean, the thundering energy like an approaching drum-beat of troubles you were destined to have but didn't know about yet.

Gunning the Jet Ski, Lickle watched over his shoulder as Hamilton released the rope. For a moment, although Hamilton was moving at least forty miles per hour, it looked as though he had come to a dead stop. Lickle watched in astonishment as the wave rose and rose and then, absurdly, rose again and rose some more, until Hamilton was in that ulti-mate place—Ant Man on the Great Pyramid of Giza—and the drop was a plunge of such dimensions that even catching the wave was a challenge. "It was all I could do," Hamilton had told me. "I was just focused on mak-ing the drop."

The singular concentration required to survive the vertical top of the wave made it impossible for him to race horizontally through the bar-rel to outrun the falling lip; the drop was so endless, with so much water rushing back up it, that he simply ran out of time. Realizing he was about to be swallowed, Hamilton deployed a last-ditch tactic to avoid being crushed: "I got as high up as I could and dove into the face." The good news was that when this maneuver worked, the surfer escaped an immi-nent beating by punching through to the backside. The bad news: he would surface directly in front of the next wave in the set. And in Hamil-ton's case, the wave that was bearing down on him measured at least eighty feet.

Lickle, as always, was on it. He raced in to pick up Hamilton and then boomeranged out at full throttle, clean grab, everything flawless. "GO GO GO GO GO!!!" Lickle recalled Hamilton screaming. The two men couldn't see what was behind them but they could feel it gaining, gaining—*WHAT THE FUCK???*—gaining. Then reality tipped side-ways, upside down, inside out, as the wave blew them off the Jet Ski in a manner that suggested extreme finality. "That wave ran us down like we were standing still," Hamilton said. "I have never been hit by anything that quick. I can only describe it as a visual of a house when an avalanche comes. You see stuff go blowing by. Well, that was what happened. We were looking at the shore and all of a sudden we were just looking at whitewater."

As the two men and the Jet Ski blasted into the air, Hamilton felt a rope wrap around his ankle. Somehow he managed to reach down and flick it off, but in doing so his shin slammed the back of the Jet Ski with such force that it cracked a baseball-size hematoma straight to the bone, almost splitting his skin. Lickle, however, was about to top that injury and raise it by an order of magnitude. Everyone at the picnic table subtly leaned forward to listen as Lickle described what happened next: "It was like being shot out of a cannon. I was blown into the sky, where I was up and over the whole whitewater—forty, fifty feet—literally flying." He chuckled darkly, no humor about it.

Behind him, Hamilton, barefoot and clad only in surf trunks, came out of the garage carrying a seventy-pound bag of pig feed. He set it down next to the barbecue. "It was a pounding," he said flatly.

"That it was," Lickle agreed. "I still get sick to my stomach talking about it." But he took the fresh beer Shearer handed him, and continued.

As Lickle landed in the tumult, he felt something hit his leg "like a sledgehammer. BOOM!" In the next instant everything went dark, as he was driven deep by the whitewater. The wave's impact rocked him so hard that when he came up, "I had no idea why I was in the water." He found himself facing shoreward, his vest holding him on the aerated surface, "but I didn't really know where I was or what I was doing." Hamilton, meanwhile, had also come up. He'd been dragged farther inside and caught sight of Lickle, seventy yards away. At that moment he also saw that the whitewater around Lickle was no longer white.

It was crimson.

But there was a more immediate problem. "I'm looking at Brett," Hamilton said, "and all of a sudden I see a *fifty-foot* whitewater behind him. Like at least. It was a whitewater as big as a lot of big waves I've ridden." Hearing this, Emory nodded, his face somber. He'd been chased and caught by a similar beast.

"It was round two," Lickle said. "Just ragged. Blasted everywhere. Head up, down, you don't know where down or up is. That second wave freakin' annihilated me." It hit with such force that Lickle and Hamilton were hurtled about five hundred yards underwater. "It was just *black* down there," Hamilton said, spitting out the word. "Pitch-black."

Once again the vests did their job and the two men popped up—just in time to face another wall of whitewater. This third wallop was then followed by a fourth and a fifth. "We had five good whalings," Hamilton recalled. "They began to blur together. Blown in, blown in, blown in. And then finally we were pushed to the inside, into the deep zone."

The two were still more than a mile offshore, still caught in wicked surf, but they'd been washed out of the worst danger and miraculously they were still near each other, only thirty yards apart. Hamilton, now freed from a survival mode in which thoughts and plans were luxuries that didn't exist, recalled the bloody water he'd seen and yelled to Lickle.

Turning his head slowly, Lickle stared at him with blank eyes. "Tourniquet," he said.

"Fuck, cuz, I was dying at that point," Lickle said, picking up the story. No one had budged from the picnic table. It was as though Lickle had just returned from Pluto and was describing the scenery. After Hamilton had assessed the injury—"Brett's leg looked like two curtains just hanging there"—and then stripped off his wet suit to make a tourniquet, he turned to what was now a life-or-death proposition: finding the Jet Ski. Fearful of leaving his friend but aware there was no other option—Emory and Denny would never find them amid the chaos— Hamilton pulled off his flotation vest and strapped it around Lickle. Then he turned and began to swim.

He spotted the Ski almost immediately—a needle-in-a-haystack sighting—but he also realized with a sinking heart that it was more than eight hundred yards away, ripping out to sea on a six-knot current. He put his head down and sprinted.

"Were you weak from losing blood?" Casil asked Lickle.

"Yeah, but I didn't go there," Lickle replied. "I just tried to keep myself together, tried not to go into the gutter. I mean as far as I knew, the Big Guy was gonna come eat me before I bled out." (Spreckelsville was, after all, the place he and Kalama had seen the fifteen-foot tiger shark.)

"At that point did you know how bad your leg was?" I asked.

Lickle nodded. "Dude, it felt like the hot jelly innards of an ahi—you know, when you're chumming with it. When you cut up a big one to use the meat."

"You *felt* it?" Somehow I found this fact more disturbing than the injury itself.

"I had to!" Lickle said. "I was dying. And I just knew something wasn't right with my leg." That something, he believed, had occurred when the thin metal fin of a tow board razored into him while they were tangled in the whitewater. "It went into the calf, straight to the bone."

While Lickle hung on, Hamilton somehow managed to chase down the Ski, fifteen minutes of hard swimming in the froth and churn, against the current. But that was only the first challenge. Lickle had been wearing the wrist lanyard—the mechanism for starting and stopping the engine—and it was irretrievably lost. Rifling through the glove compartment, Hamilton found a pair of iPod headphones that he used, MacGyver-like, to hot-wire the ignition. To his relief, the battered machine started up immediately. As he gunned back to Lickle he radioed a coast guard alert; he wanted to make sure an ambulance would be waiting at Baldwin Beach.

He found Lickle semiconscious, in shock, and floating in a pool of blood, but he was still alive. Hamilton managed to get him onto the rescue sled in a sort of half-kneeling position. Holding Lickle in an armlock so he didn't slide off, Hamilton headed for shore.

"When Laird shot up onto Baldwin Beach," Lickle said, "he was stark naked. By then I had lost most of my blood. I had no idea that he had pulled his whole wet suit off to tie the tourniquet. Even when I was on the Ski and my face was in his ass, I still never put anything together."

The swell had sent waves washing through the parking lot, two hundred yards from the shoreline. A beach rescue vehicle had been swept up there and the lifeguards were milling around, Hamilton recalled, "kind of flipped out." As he tried to lift Lickle off the Jet Ski, the ambulance pulled into the flooded lot, then the fire department, and finally the police. "I had to scream at one guy," Hamilton said. "He saw the wound and it tweaked him." While the paramedics attended Lickle, someone handed Hamilton a T-shirt to cover himself. Someone else passed over a pair of surf shorts. As all this was going on, Shearer and Kalama pulled up. Only moments later Emory and Denny arrived back on shore.

"Put the shorts on," Hamilton said, "made sure Brett was in the ambulance and everything was okay and then we went, you know, 'Okay, back at it.' "

"Wait a minute," I said, wanting to make sure I understood correctly, what with the vague pronouns and all. *"You went back out?"*

Before he could answer, Shearer cut in. "Dave and I got to the beach and the ambulance was just leaving. And Laird's eyes are like friggin' lightbulbs. I mean he's jacked beyond jacked. I've never seen him more jacked—and we've been through a lot. He can barely walk; his shin is all blown up. But he's just so on a mission. And he looks at me and says: 'You have to see it.' "

"What do you mean!" I stared at Shearer, startled by this information. Until that moment I'd felt a kind of semidetached third-party horror about December 3, a rubbernecker's fascination. But now something new burst in: jealousy. *"You were out there?* You saw those waves too?"

Shearer nodded vigorously and leaned forward for emphasis. "Have you ever seen *The Poseidon Adventure?* Well, this was my personal Poseidon Adventure. Times *ten.*"

Emory headed home for dinner and Gabby called Hamilton upstairs for a moment and Shearer kept on with his story. He had gone back out with Hamilton, and Emory—now partner-less since Denny had stayed on shore—had accompanied them, driving the other Ski. "So I'm holding on for dear life," Shearer said, "and we couldn't really see, there was so much salt spray and moisture in the air." He leaned back at the table. "I mean, you had to *really* want to get out. We had to continually attack and retreat, attack and retreat. We'd see what we thought was an opening and then we'd have to run away. Just to get through all the whitewater." By that time the entire coast was a fifty-foot closeout: "The foam was four feet thick!" "At one point I got knocked off the Ski," Shearer recalled. "Laird grabbed me and said, 'Where do you think you're going?' Then we started getting glimpses of what was out there. *Unreal.*"

"Wasn't Lake Havasu, was it?" Lickle said, with a smirk.

"Listen," Shearer shot back. "I'm used to looking at stuff. I've seen every big swell in Hawaii since 1986. I've filmed almost everything that's ever happened as far as surfing goes. I've been first responder at a plane crash with twenty dead bodies. But as we tried to get out, I was just totally tripping." He looked at me. "Finally, we got out. And it was like, 'Oh. My. God.'"

"So how big *was* it?" I asked again.

Shearer paused. The wind had taken a break for a moment and his silence seemed bigger. He fixed me with a searing look, straight in the eye. "I fly every day with a hundred-foot line on my helicopter," he said. "I can tell you when it's inches above the ground or a foot above the ground or two feet above the ground. I can tell the difference between a hundred-foot cable when I'm flying or a hundred-five-foot cable or a ninety-five-foot cable. I am very good at judging height. I have to be." He continued, his voice ratcheting louder. "And I know for a fact—I

KNOW—it was over a hundred feet out there. GUARANTEED. I'm even saying one-ten, one-fifteen. And I would go out on a limb and say that some of those waves were one-twenty."

Lickle nodded. "I call it eighty because I like to not, you know, stir people up, but it was fifty-foot Hawaiian." He clarified: "There were hundred-foot waves rolling through that place."

Shearer stood up and held out his arm, covered in goose bumps. "Look, brah: chicken skin." He sat down again, and reached for his beer. "All I know is what I saw."

At Egypt, here was what greeted his eyes next: Hamilton, his left knee and shin swollen twice the size of his right, had picked up his board and was reaching for the tow rope that Emory had tossed into the water. "I gotta get one more," he explained. Bailing off a wave and almost losing his partner was, in other words, not the way Hamilton intended to finish out December 3: "I can't leave here whipped." Shearer, then, found himself solo on the other Ski. Even for a guy whose résumé includes flying into a live volcano to save the people aboard a downed sightseeing chopper from perishing in a caldera of toxic fumes and molten lava—despite keeping his cool in situations like that as a matter of course—looking at Egypt's waves, Shearer was scared. "But I said, 'I'll be fine, brah. I'll just stay way outside.'"

Desperate not to get caught inside the breaking giants but not altogether sure where the giants were breaking, Shearer, who had operated a Jet Ski all of three times before this, proceeded on high alert, driving up forty-degree shoulders as the waves were still building. It was the safest thing to do, if *safe* was a word that could possibly be applied to any part of the situation. But by staying so far on the shoulder, Shearer realized, he had no chance of actually watching Hamilton ride a wave. Loosening the death grip he had on the wheel, Shearer gathered his courage: "I told myself, 'I know I can do this. I know where to be. I know I can do this.'" After a few rounds of incantations, he caught sight of Emory towing Hamilton onto a second wave—and he began to follow them. "I was on

Sierra's far right," Shearer said, "and he was going for it. I had a side view and a quartering left view of the wave; I could see to the back and I could see the reef in front. I watched Laird release the rope and then I saw him come down this wave . . ." His voice trailed off and he focused his eyes on the pineapple fields, fighting back emotion.

After a moment he inhaled briskly and continued. "I've never seen anything like it in my life. The whole reef drained! And the energy that the wave had taken to jack itself up created this trough. It was like . . . the bottom of the wave was ten feet below sea level! And Laird's on the face and the reef is drained off and there's this THING behind him. And I saw it! I saw it. *I saw the Big Mother.*"

I stayed in Maui, digesting what I had learned. By this time in my reporting I'd been spitting distance from sixty- and seventy-foot faces, and Teahupoo's grinding barrel, but now I understood that the waves on December 3 had been something altogether different. As the story came into focus, I felt overwhelmed. I wanted to be able to envision the day's sweeping vistas and tiniest details, down to what the air smelled like. Past experiences had taught me that strange energies can arise in the bull's-eye center of a storm. Ions zip around, frantically changing polarities, and when the barometer plunges, that pressure change affects the water in potent ways. Memorable violence can spring up in a hurry.

One August at my family's summer cottage in Canada, my father and I had been caught out on the lake when a tornado descended. I was swimming from one side of the lake to the other. He was driving shotgun to make sure I didn't meet up with another boat's propeller. His golden retriever, Bear, surveyed the scene with concern, paws draped over the gunwale. When I'd dived in an hour earlier, the water had been agitated and choppy, its green-black depths illuminated by flashes of sun that poked through the clouds. The afternoon was pretty, kind of lively. In minutes all of that changed. It was as though a great vacuum had come along and sucked the light and breath out of the water and sky. Everything became dark and still. "Get in the boat!" my father shouted, and as he did, a long low breeze swept the lake's surface like a chill along its spine. The dock was only four hundred yards away, so instead of climbing into the boat I sprinted for land. As the tornado closed in, the waves rose and I became aware of a smell, an odd, moist, chemical scent borne on a current of static electricity. Out of fear I'd been swimming with my eyes shut, but for some reason I opened them as I neared the ladder.

Through my goggles I caught the stealth silhouette of a loon streaking below me. The water was black and the loon was black, but tiny air bubbles streamed off its wings so I could make out its shape and that of a baby riding along on its back, and as they passed beneath me the baby loon twisted its long neck and looked up with glittering red eyes. If I wasn't rattled then, I was ten minutes later inside the cottage when the tornado hit, uprooting fifty-foot pines, shattering windows, tearing off shingles, flicking cars off the roads, downing power lines, reminding everyone of their essential flimsiness. The electricity stayed out for a week. But despite all the damage what I remembered most was the storm's eerie vibe—an observation echoed in the stories of December 3—and how an average day had suddenly turned into something truly malevolent.

Later that week Jaws broke. The waves were twenty and thirty feet tall, too small for towing but a challenge for standup paddle surfing, so Hamilton went out, taking Ekolu Kalama with him. It was an evening session. Teddy Casil and I drove to the cliff to watch, the Mule bumping over red dirt paths through silver-green fields. The air was soft, no crisp edges. Clouds lined the horizon, lavender, peach, cornflower, and gold, and the ocean gleamed a six-dimensional navy blue, whitewater spilling to the cliff as the waves broke. Offshore, Jaws' crest flashed like an enormous smile. The only thing missing from the scene, I thought, was a pod of frolicking dolphins.

We pulled up to the lookout next to Lickle's golf cart, red with orange flames emblazoned on its flanks. Lickle stood at the cliff brandishing a two-way radio, deep in conversation with Emory. A handful of spectators bunched along the edge; one guy was setting up some camera gear. I walked over to Lickle and Emory, catching the tail end of something Lickle was expressing in a vehement voice: " . . . because I'll never BE there again. I'm not going to be out there on the next Day of Days. I am not. I have no desire whatsoever."

I'd heard talk of this, that Lickle's accident had effectively retired him, and I'd wondered if it was true, or even possible, given that riding giant waves seemed to be more of a calling than a choice. "Really?" I asked, wedging myself into the conversation. "That's it? You're done?"

"For that, yeah," Lickle said, and then added: "I am not over riding. I'm just over IT. I'm over the quest for biggest friggin' waves that man can ride." He paused for a second, his eyes scanning the lineup at Jaws. "Because where does it stop? Do I need to be the guy who's towing Laird into Cortes Bank because they've got a hundred-twenty-foot swell coming? No."

"Well, you want to surf a wave," Emory said. "Not just survive it."

"That explains where I am," Lickle said. "Survival takes all the fun out of it—unless you're Larry." He smiled. "Of course, there's a point where he's surviving too. We were at a whole other level that day—the level where even Laird gets scared, you know what I mean? So all of a sudden the things you take for granted from him, you can't take for granted anymore. At eighty feet he's starting to feel the same emotional pressures we feel at forty." He turned and spoke into the radio: "Okay, two behind this one, Laird." (Hamilton would hear him, even out in the waves, through the waterproof radio he kept clipped to his surf trunks. With no Jet Skis accompanying him, the radios were a safety measure more than anything, but it didn't hurt to take advantage of Lickle's hundred-mile view either, as the swell pumped across the horizon.) I watched the incoming set, trying to figure out how Lickle was calling it, wondering if I could discern why wave number three would be the most desirable. Maybe my wave-reading skills were improving: I thought I could pick out one shadow that was slightly more pronounced than the others.

I fixed my binoculars on the wave. Paddling in from the side, Hamilton and Kalama showed up as miniature silhouettes in the golden light. Jaws loomed in front of them and then Hamilton began to paddle furiously, leaning forward and digging in the blade with everything he had. Beneath his feet the ocean rose until the wave was ready to break and then Hamilton dropped onto the face, touching his paddle lightly to the surface to make minute balance adjustments. The wave arched above him, but just before the lip closed Hamilton shot out, flirting with being swept into the whitewater but managing to stay just ahead of it.

"Oh, he's running the rocks," Lickle said with pride in his voice, watching Hamilton make for the jagged boulders jutting up from the water, weaving through them all the way to the inside. Lickle turned to me. "Just so you know, that was very dangerous."

"I'm sure after December third it doesn't feel that way," I said.

"Listen," Lickle said. "One of the worst hold-downs I've ever seen here was on a ten-foot day. It's like, when you drop your guard and you think, 'Ah, this one's not that big,' or you act defiant . . ."

He clicked on the radio: "Come back, nice waves, Laird." Out on the skyline the whole ocean seemed to be undulating. Lickle continued: "Plus, he's wearing a leash and no vest. Fall, and you're gonna do some time. If I wasn't wearing a vest during my ordeal I wouldn't be alive right now."

"Telling war stories? Ninety feet and glassy is what I heard." The photographer, a well-known local named Erik Aeder, had walked over to join us. Though the Maui papers had briefly covered the incident, everyone was curious to know more. Aeder shot a glance at Lickle's scar.

"It doesn't get any bigger or any better, I can promise you that," Lickle replied.

"That was another one of your nine lives," I said, trying to be light about it.

"And I've got seven or eight of 'em already used up," Lickle said. "Every time it happens I just go, 'Wow, man, someone loves me. The Big Guy loves me." He smiled. "I'm baffled that God loves me to the level that he does."

"You've got work to do here, obviously."

"I know that." Lickle reached down and picked up a plastic bottle that someone had thrown in the dirt. He tossed it into the back of his golf cart. "For that board to have hit me the way it did and not killed me, that was destiny. There's something I'm supposed to do. I don't know what it is yet, but there's a reason I'm still here. And that's a cool feeling."

"What I still can't believe," Casil said, "is that you could hold your breath for that long."

"I *can't*." Lickle shook his head. "That's the whole point. You can do a lot of things you don't know you can do. Once you're in survival mode there's no skill. It's instinctual. And we all have that."

"A lifetime of preparation doesn't hurt though," I said.

"Yeah, but you know, I smoke pot, and I'm not a breath-holding fanatic or anything," Lickle said, referring to the training that some big-wave surfers did to increase their lung capacity. (Hard-core practitioners run underwater for minutes at a time carrying twenty-pound rocks.) "But I do know there are breathing thresholds you can go through. You hold your breath and you hold it and then all of a sudden you'll feel this tensing, this panicking, and then it releases. And then you'll have a whole other period before the next panic. On the first wave I went through about three or four of them. You can get five if you're lucky."

"What I still can't believe is that you and Laird went back out." I aimed the statement at Emory.

"Yeahhhh," Emory said, drawing the word out slowly. "I went back out to support them. It was better we had two Skis out there. I didn't even want to catch any waves at that point. It was a freaky feeling. It was big and scary and I didn't need any of it. And it had gotten even bigger! I'd already caught the biggest wave I've ever been on, and I'm like, 'No thanks, that's good. I'm still alive.' Having Don out there on the second Ski and no one around . . . It just didn't feel safe. That left break was over hundred-foot faces. No question."

Lickle nodded in an exaggerated way, as if to say, *No shit.* "I couldn't believe Laird put Don on the Ski," he said. "Because you have to ask any-one who's been out there: Can you take it when it goes down? You are in the middle of the friggin' ocean. And you have *nothing.* You are at the mercy." Noting the darkening sky, he gave the surfers a heads-up: "Ten-four, Laird. You only got so much light left."

Hamilton responded on a crackly connection: "Little tricky on the inside here. It's Ekolu's first time so we don't want to baptize him too hard."

"Roger that," Lickle said, laughing. "But he did catch a wave so he's been baptized. You'd better let him know it's time to go. There's a serious set coming in."

"Yeah, we're going," Hamilton said, signing off. "Okay."

The two men, alone on an expanse of ocean that stretched as far as the eye could see, began to paddle in the direction of the Old Fishing Shack, from where they had launched. The sun slipped below the water, turning the sea to mercury. The waves continued to flow in, like music that had arrived here across an endless continent of water, some waves faint and some loud, all eventually coming to a crescendo on the reef. As they did so, their soft roar was hypnotic. It was impossible to stand at the cliff and feel anything but gratitude. No wonder the Hawaiians had per-formed their most sacred ceremonies on this bluff. The air was thick with their spirits.

Down below, Hamilton and Kalama suddenly changed direction, U-turning back to the wave. They'd spotted something they couldn't stand to paddle away from. Casil handed me the binoculars. "Check it out," he said, pointing to the north. Galloping in, swinging toward the reef at a slightly cockeyed angle, a foreboding lump rose from below as though a colossal hidden object—an island, maybe, or a ski hill—had decided to surface.

"Oh, look at that set!" Emory said, as Kalama began to race for a wave. "Oh, Ekolu! Turn, brah! Paddle! Paddle! Paddle!"

"There he goes," Lickle said, approvingly. "He's gonna make it! He wants it! But—oh—baby, look at number four behind it!" As he said it, I saw Hamilton quickly reposition himself. Even in the dusk he had seen the outlier too, or maybe he had felt it in some primordial corner of his senses, the same way the birds at Lituya Bay somehow knew in DNA Morse code: Big. Wave. Coming.

"Whoa!" Emory said, looking at Hamilton on the wave. "Did you see that? Laird leaned forward. He mispaddled. He almost fell!"

"You know what, though?" Lickle said, leaning back against the golf cart and smiling. "He didn't."

I'VE SEEN THREE OF MY CLOSEST FRIENDS KILLED BY WAVES.
ALMOST BEEN THERE MYSELF.

Big-wave surfer Mike Parsons

OUT, WAY OUT, ON THE CORTES BANK

ANAHEIM, CALIFORNIA

On a sultry spring evening in the long shadow of Disneyland, deep in the neon heart of southern California's Orange County, the crowds began to arrive at The Grove, a cavernous, tan stucco event space. The building was designed to invoke a golden-age Hollywood movie studio (down to the fake water tower), but instead of the MGM lion or Columbia's torch-bearing goddess serving as resident icons, on this night The Grove was festooned with banners and video screens sporting the Billabong surf company's XXL insignia, and the dripping-lime-green M of Monster Energy Drinks. Instead of Errol Flynn and Ava Gardner strolling the grounds there were throngs of big-wave surfers and their sponsors and their girlfriends and their fans; surf photographers and surf glitterati and surf hangers-on. In all, more than two thousand people were expected on this night for the Eighth Annual Billabong XXL Global Big Wave Awards, described by its founder, Bill Sharp, as the sport's "Oscars." More than $130,000 in prize money would be doled out, not to mention bragging rights for the year's Biggest Wave, Monster Paddle, Biggest Tube, Best Wipeout, and the grand prize: Ride of the Year.

Eight years after Sharp (with Billabong's backing) had launched The Odyssey: The Search for the 100-Foot Wave, the event had been renamed, reconfigured, and toned down, less specifically focused on that three-digit grail. In fact, the XXL's Biggest Wave category now paid out only $15,000 to the winning surfer—along with a new Honda Jet Ski, with an additional $4,000 going to the photographer who shot the ride— cash worth visiting Anaheim for, to be sure, but a far cry from the $500,000 Sharp had originally dangled.

There were a number of reasons for the changes: cost, liability issues, and logistical hurdles among them. In July 2001 Sharp had announced the Odyssey with fanfare, touting it as the "ultimate man-against-the-sea adventure" and saying things like "We'll invest in the television production to make sure that when the one-hundred-foot wave goes down, it's shot perfectly from every angle." His original vision involved shuttling an elite team of riders around the world wherever and whenever a hulking magenta blob popped up, hunting down mysterious, unridden waves in the far corners of the world's oceans. To that end, Bill-abong had even purchased an amphibious plane called the *Clipper*, designed to land in rough seas. "For the first time ever," the press release read, "a band of surf adventurers will have at their disposal a way to move faster than the weather, unrestricted by scheduled airline flights, paved runways, or even paved roads."

In interviews Sharp had spoken excitedly about breakthroughs that his competition would bring to tow surfing, establishing new protocols for the sport: "We will, over the course of this entire project, be develop-ing new equipment and new procedures that will make the tow-surfing experience safer and more enjoyable." What kind of new equipment? "We're looking at little miniature air tanks, GPS locators, or whatever it's going to be," he told *TransWorld SURF Business*. "We're going to be delving heavily into the James Bond gadget arena. Maybe there'll be a mad scientist like Q in the back of the Billabong warehouse creating secret gizmos."

Such grandiose plans delighted many big-wave surfers and rankled others. Hamilton, who had spent more time thinking about tow-surfing gear and rescue procedures than anyone (not to mention stunt-doubled for 007 himself, surfing sixty-foot Jaws in *Die Another Day*), was foremost among the latter group. To him Sharp's vision amounted to a craven media play made for all the wrong reasons. "This one-hundred-foot-wave thing," he said. "I resent it. The intentions were never truly genuine. It was always about 'How can I exploit this?' and 'How can we maximize our marketing dollars?'"

Hamilton's irritations were compounded by the fact that, in the contest's early days, far from scouring the world for new waves, whenever a big swell came along, everyone headed straight for the sure thing: Jaws. "In 2002," Hamilton recalled, grimacing, "they all came out of the woodwork. There was more wiping out in one day than there had been in ten years. I watched a guy lose two brand-new Jet Skis in less than five minutes. After that, on any big swell we were like, 'Hey, the crash test dummies are here!' They've pulled up a truck full of them and they've sent 'em out. *Crash, crash, crash*—'send the next ones in!' *Crash, crash, crash.* 'Okay, send the next ones!'"

As hordes of marginally qualified bounty seekers poured into Jaws, tempers frayed to the point where Dave Kalama, encountering Sharp at an event called the Waterman's Ball, expressed his frustration physically. "It was at the time when he was claiming everything, like 'I'm the big-wave blah blah blah,'" Kalama said, describing the incident. "I saw him at the bar and I went, 'Hey Bill, come here. I want to talk to you.' So he walks over and he goes, 'Oh, hey Dave, how's it going?' and I said, 'Good.'" At which point Kalama reached for Sharp's outstretched hand and instead of shaking it, folded him into a headlock. "And I said, 'You represent everything I hate about big-wave surfing.'"

Things had mellowed in recent years, most likely due to the fact that there hadn't been any big days at Jaws for people to fight about. Sharp had adopted a lower profile, making fewer hyperbolic claims. But hell would still have to sprout icicles before Hamilton, Kalama, Lickle, Emory, or Doerner would ever show up on the XXL red carpet. "I'd rather clean toilets. I'd rather step on a nail—a big, fat, rusty one—and then have to get the fifteen-year tetanus shot," Hamilton said, listing things he would rather do than stroll down Billabong's floodlighted stretch of velvet rope lined with photo blowups of the year's nominated waves; occupied on this night by clusters of men in surfer black tie—dark jeans, dark T-shirts, the occasional sport jacket—flanked by young women in as little as possible.

It was a surf-industry night out, a convivial frat house where the mark of inclusion was to be known by your nickname. "Yo, Rippy!" one attendee yelled. "Shasta—wassup!" Many of those strolling into the theater wore sunglasses, even as darkness settled and the royal palms that flanked the entrance became noble silhouettes. Camera crews roamed, documenting the event for an eventual ESPN2 program and a live webcast. Bill Sharp was omnipresent, clad in head-to-toe black, his tall frame and spiky white hair making him easily visible as he circulated.

One of the red-carpet interviewers, a striking blonde in her twenties, held out her microphone to a young man in wraparound shades. She welcomed him to the event and asked his name. "My name is Rat," he said, revealing with his first words a state of deep inebriation. "I just got out of jail," he added in a celebratory tone, "and I'm totally good." The interviewer, looking flummoxed, turned away, while the cameraman suddenly cut to a trash can. "Hey, get your ass back here!" Rat yelled as they retreated.

By eight p.m. most of the nominees had emerged from the pink and black Billabong tour bus parked out front and could be seen milling around beside their photographs, doing video interviews. A delighted-looking twenty-one-year-old Australian named Mikey Brennan, nominated for both Monster Tube and Ride of the Year, talked about his home wave, Shipsterns Bluff, a seething slab off the Tasmanian coast, with a Billabong interviewer nicknamed G.T. "I live there, so I'm always there," Brennan explained. "It's nice to surf there, especially when there's waves."

"Well, trust me, mate," G.T. informed him, spotlights glinting off his sunglasses. "There are a few girls here who will want to meet you, especially if you win."

You had to root for Brennan, with his tousled mop of hair and mouthful of braces, but he had some serious competition: Shane Dorian, fresh off the plane from Fiji; Ian Walsh, in from Maui; and Tahitian star Manoa Drollet. All three men were up against Brennan for their performances at Teahupoo the previous October. Brazilian rider Carlos Burle was the fifth finalist in the Ride of the Year category, for a wave at Ghost Tree, on December 4.

A few yards farther along the red carpet, Garrett McNamara stood surrounded by well-wishers. "I'm on my way for a monthlong mission somewhere," he said and, when asked for specifics, responded: "That information can't be disclosed yet." Tonight McNamara was nominated for the season's Best Overall Performance, and also Wipeout of the Year, for a head drop he'd taken at Mavericks. Right behind him, Brad Gerlach described his December 4 experience at Ghost Tree for a video crew. "It was foggy and weird," he said. "But God, what an exciting wave. It's fun because it's so scary." He laughed. "*Well*, it's more fun when you're done riding." In the background, clusters of surf fans drank beer and cans of citrus-flavored Monster, spiked with vodka.

Looking at the posters of the nominees, it was clear that the 2007–8 big-wave season had produced an excellent crop. Giant waves had raged from Tahiti to the Basque Country, from Oregon to Mexico to Ireland to Spain to Tasmania to South Africa. Hawaii, as always, had its moments. California was surreal. "The first five days of December were possibly the best five days in the history of big-wave surfing," Sharp proclaimed. This year's winners had been chosen from about five hundred entries, each of them scrutinized by judges who used protractors and other measuring tools to gauge the waves' height.

As the crowd began to filter inside for the ceremony, G.T. and his cameraman stopped a woman with blond hair rippling down her back, wearing a black minidress. This was Maya Gabeira, a twenty-one-year-old Brazilian and the leading contender for the dismayingly named Girl's Best Performance award. This year Gabeira had ridden all the biggest swells, turning heads with her beauty and her dauntless pursuit of waves that terrified most male tow surfers.

"You've been a busy little lady," G.T. said greasily. "Are you single?"

Gabeira graciously ignored him and began to speak about her season in a lilting Brazilian accent. "I spent most of the winter between Hawaii and California. Then I—" G.T. cut her off. "You're so hot," he said. His hands swam through the air, gesturing at her. "How can you be so feminine and ride big waves?" Gabeira flinched backward and disappeared into the crowd as quickly as possible.

Delivery aside, the question was an interesting one. While the women's professional surfing tour was filled with contenders, there was an undeniable shortage of female tow surfers. In fact, as a team rider for both Billabong and Red Bull, Gabeira was one of a kind. Her two fellow big-wave nominees, Jamilah Star and Jenny Useldinger, both with talent and guts to spare, focused on paddle surfing rather than towing. Neither woman had attained Gabeira's level of sponsorship.

I'd met Gabeira in Tahiti, where I watched her take two life-threatening wipeouts on Teahupoo, dust herself off, and head right back out there. Towed by Raimana Van Bastolaer and her mentor, Carlos Burle, with Hamilton patrolling the sidelines, she went on to ride four waves in the same conditions that had also gotten Dorian, McNamara, Walsh, and Drollet nominated tonight.

As far as big-wave careers go, Gabeira's had followed a steep and improbable trajectory. When she learned to surf at the (relatively advanced) age of fourteen: "I was scared of one-foot waves. It was challenging just to be in the ocean. So when I accepted the fact that I was able to surf one-foot waves, I went to two-foot waves and then three-foot waves. I've kept going until now." There were dozens of reasons why she might not have succeeded, but Gabeira had at least one advantage: as the daughter of Fernando Gabeira, a famous dissident against Brazil's military dictatorship in the 1960s and 1970s who went on to become a crusading politician, she had the right genes. "Yeah, my dad's crazy," she told me. "He's a tough man. Like really, really tough."

Despite the accolades that were flooding in, Gabeira knew she had a long way to go before she achieved her goal of being one of the best tow surfers—male or female. She wanted to master Jet Ski driving and rescue techniques. She needed to get out in even larger waves. And she had to reckon with the limitations of the XX chromosome when it came to such intractable things as body mass. "Even as much as I train I'm still not that strong," she said. "Big waves are already aggressive if you're paddling in, but when you're in there with a machine like a Jet Ski—it goes so fast and it's so heavy—that just makes it in reality a man's sport. There's a lot of impact. My body suffers when I do a tow session even in small waves."

Determined to change that, Gabeira had recently added Bikram yoga, weights, and mountain biking to her endless surf practice. In the future, she hoped, her example would bring more women to tow surfing. "I don't know how many years it will take to change," she said, sounding slightly wistful. "It's hard, you know, being the only girl. You get a little bit intimidated. The whole lineup, the wave, is dominated by men. You think, 'Hmmm, can I really do it?' "

Inside the theater a video montage of the year's noteworthy rides commenced on a giant screen above the stage, forty-five minutes of surfers on megawaves and beach girls in thong bikinis set to a heavy metal beat. People were seated at long tables, a type of supper club arrangement that enabled The Grove to serve its audiences drinks throughout the entire evening. Blue and red lights played in the darkness, and everywhere you looked there was a logo: an XXL stamp or a floating, disembodied M or a nod to Verizon or Surfline or Hawaiian Airlines.

The video wrapped up and the event's host, Australian surfer Mark Occhilupo, stepped onto the stage. Compact, with shoulder-length, sandy brown hair and a sly smile, the 1999 world champion was known affectionately as Occy. Standing at the podium, he drew loud, heartfelt cheers. "What a year!" he began in a thick Aussie accent. "This thing's getting so big. We're getting entries from all over the planet." Occy faltered, looking down for a moment. "Um, I just want to say a quick thing with Peter Davi. He was a good friend of mine. I'll miss him. He was so cool." A subdued cheer rose up, a collective *shaka*. Many of the attendees had known Davi too, and as a big-wave surfer who had died in the ocean his name was now inscribed into the sport's sad pantheon of casualties. "Apart from that," Occy veered abruptly, "we're gonna move on to the Monster Tube award. It's definitely one of my favorite experiences, tube riding. But these guys are fishing with longer poles."

Presenting the award was an Australian snowboard announcer known as Dingo, and a pro skateboarder named Rob Dyrdek. Dressed in a green hooded parka, a scarf, a Monster baseball cap, and a huge pair of sunglasses with powder-blue frames, Dingo screamed into the microphone, *"You guys are the craziest motherfuckers I ever met!"* before opening the envelope and declaring Manoa Drollet the winner. Drollet came up onstage, greeted by two hostesses in miniskirts, stiletto heels, and wraparound smiles who handed him a giant cardboard check for $5,000.

The festivities rolled on. Wipeout of the Year was awarded and the prize, a Samsung phone, seemed cruelly insubstantial. Gabeira, to no one's surprise, won her category for the third year in a row. ("Congratulations, Maya," Occy said as she left the stage. "And grab me a beer, will you?") Greg Long took the Monster Paddle Wave award for a fifty-one-footer at Todos Santos; he thanked his friends and family for putting up with his "obsessive-compulsive behavior chasing these swells" and apologized to his mom for choosing a big day at Mavericks over Thanksgiving dinner. They didn't come any more charismatic or talented than Long; only twenty-four, he was already at the top of the game—and still rising. Ten minutes later he also won the Best Overall Performance award.

The hooting grew louder as the Biggest Wave finalists flashed up on the screen. Ghost Tree had produced two: Gerlach's ride, and that of another rider named Tyler Fox. Over in France, a wave called Belharra Reef had made surfer Vincent Lartizen a contender. But it was the last two nominations that caused everyone to lean forward in their seats. Both rides had taken place at the Cortes Bank on January 5, 2008, under haunting conditions. Only a handful of photographs existed of these waves and the surfers on them, South African star Twiggy Baker and big-wave veteran Mike Parsons. Not a single frame of video had made it back from the session. When you heard the story of how that day had gone down, you could understand why.

"I knew it was going to be huge, but I didn't expect it to be *quite* that big." Mike Parsons looked at me across the table of a quaint San Clemente café and took a bite of his sandwich. His face was distinctively angular, his eyes a piercing pale blue set off by his tan. He was tallish and lean, casual in a Billabong T-shirt and jeans. But the most noteworthy thing about Parsons's presence was that despite his many triumphs, his career longevity, and the respect he commanded in his sport—he gave off a vibe that could only be described as gentle. Experienced tow surfers stressed the importance of humility when faced with the waves' overwhelming power, but Parsons exuded it even on land. There was zero flash and bluster about him.

"They were definitely the biggest waves I've ever seen," he continued. "The ocean was just so alive. I mean, we knew it would look spectacular, but the playing field and the amount of chaos and commotion on that reef was amazing. I think there was probably a mile and a half of whitewater out there."

The day Parsons was describing, January 5, 2008, was destined to go down in big-wave history as a seat-of-the-pants moon shot in which—somehow—everything went right. Parsons, along with Brad Gerlach, Greg Long, Twiggy Baker, and photographer Rob Brown, had braved one of the ugliest Pacific storms in years, one during which West Coast residents were wary of even leaving their houses, to head out to the sub-merged mountain range known as the Cortes Bank, one hundred miles offshore from San Diego. After driving a small boat and a Jet Ski for six hours through squalls and heaving seas, they'd spent five spooky, exhila-rating hours alone on the bank, ridden seventy- and eighty-foot waves and seen hundred-footers breaking farther up the reef, tantalizingly close but not properly situated to ride. Then, as night fell, they had turned around and spent six more hours gunning home in the dark with another Pacific storm breathing down their necks. When I'd learned about this expedition I was anxious to talk to the men, so I drove to San Clemente, Parsons's home base, to hear about it in person.

Back in 2001 Parsons and Gerlach had been the first tow team to ride Cortes in extreme conditions. For years various wave aficionados, includ-ing Sean Collins and Bill Sharp, had kept an eye on the place after being tipped off by Larry Moore, the former photo editor of *Surfing Magazine*, who in turn had heard fishermen marveling at the huge waves they'd seen out there. Throughout the 1980s Moore's curiosity built to the point where in January 1990, when a promising swell arrived, he'd chartered a plane to fly out and look around. Upon arrival he saw perfect, deep-bellied, fifty-foot waves; 100 percent rideable and utterly unlikely, given that they appeared to be breaking in the middle of the Pacific Ocean.

Below the surface, however, many factors contributed to the exis-tence of what Collins referred to as "one of the Seven Wonders of the World if you're a surfer." From the Pacific seafloor five thousand feet down, the underwater mountains that form the bank rise to within six feet of the surface at a spot called Bishop's Rock. It's on the same geo-logical spine as the Channel Islands, farther north: Cortes was simply another island in the chain until a few thousand years ago, when the sea level crept higher, slowly drowning it. Emerging from the depths, it acts as one long trip wire for swells that have stampeded down from Alaska, focusing wave energy onto the bank like a giant magnifying glass. This refraction is so dramatic that when a swell hits Cortes under the right conditions, it can jack up to quadruple its size (meaning that a twenty-foot swell can churn out eighty-foot waves). The bank's unique location, surrounded by abyssal waters and with nothing to buffer it from the full force of the open Pacific, made it a top candidate to produce not only a clean one-hundred-foot wave but, according to Collins, "definitely a very good, rideable 150-footer."

Surfers described the spot as creepy and otherworldly, and yet it exuded an irresistible pull. "There's a lot to do out at Cortes," Hamilton had said, noting its potential. The previous summer he had arranged to have access to an oceangoing jet boat that was moored in Malibu, and I knew he was just waiting for the right time to go. Problem is, the ideal conditions for Cortes came along only once in a blue moon. "You really need a stable environment out there," Collins explained. "That's a very hard thing to get for a spot that's so exposed. You don't want to be riding a sixty-foot wave and have a thirty-foot wave coming at you from another direction."

The submerged bank was twenty miles long—on par with Catalina Island—and with such a wide span of swell and wind directions there were countless wave scenarios. "There are three different takeoff spots depending on the day," Collins said. "You've got swells circling around the reef everywhere and currents going in different directions." Even if conditions looked pristine to a rider examining buoy readings and weather predictions from shore, one hundred miles out to sea anything could be happening: thick fog, banshee wind, dead calm, sudden squalls. Of the ten times Parsons and Gerlach had made the voyage to Cortes since 2001, they'd scored big on only three occasions. "Once you get past San Clemente Island everything changes," Parsons said. "It's almost as though Cortes has its own climate system."

Further complicating matters, no landmarks existed for the riders to gauge their position; in the lineup there were no lineups to be had. "Every other surf spot you're looking at land," Parsons said. "It's so weird that there's nothing but you and the ocean. It's like a spiritual deal."

Of course there was a daunting aspect to this total isolation. When something went wrong at Jaws or Mavericks or Ghost Tree, the hospital was a ten-minute helicopter flight away. At Cortes Bank a rider was hours from land, and that's if he was lucky. Aerial support was far from a given; single-engine aircraft don't have the fuel capacity to stay one hundred miles offshore for long, and in rough conditions they can't fly out there at all. If a rider fell on a big day and his partner lost sight of him, there would be no overhead spotter to pick out the tiny head in the churning whitewater and colliding currents, and no way for a Jet Ski to survey the vast expanse of ocean. There were also parts of the bank that were such turbulent impact zones that they could not be entered. "You could easily—*easily*—be washed out and never seen again," Parsons said.

Given the difficulty of getting out to Cortes and the dangers that awaited you when you did, the trip was a serious proposition even in fine weather, and the first week of 2008 had been the opposite of that. Three brutish Aleutian storms were stomping toward the West Coast; the first hit on January 4 with hurricane fury, announcing itself with 150-mile-per-hour winds, flash floods, and mudslides, toppling power lines and flipping eighteen-wheelers, shutting airports, hurling trees across roads, and burying ski resorts under ten feet of snow, killing at least twelve people. And that was only what happened on land. At sea, a merger of cold fronts and subtropical moisture whipped the North Pacific into a frenzy. Harbors were closed from British Columbia to Baja. High surf warnings were issued.

"It's a wasted swell," Collins said when I called him on January 3 to see if anyone was going anywhere to try to ride anything. There was no high-pressure ridge holding the worst weather offshore, he explained. "This storm's gonna roll right on through and bash the coast. It's going to happen right on top of us." There would definitely be giant waves, "but they'll be all sloppy and weird." This was a tempest to sit out entirely, it appeared, and staying onshore did not seem voluntary: the coast guard had raised its storm flag, signifying peril to anyone who ventured into the ocean.

Down in San Clemente, Mike Parsons and Greg Long hunched over their computers, poring over weather data as rain thundered down and gale winds rattled their windows. Despite the dire forecasts, they believed there might be a sliver of calm between the first storm on January 4 and the second, expected to hit in the wee hours of January 6, during which they could make a stealth dash to Cortes. If they were lucky, there would be a half-day gap between the retreat of one cold front and the advance of another. That respite, if it occurred, would contain all of the waves and none of the ruining winds. It was a dicey call, however, because no one could guarantee the speed at which the storms were moving. If the second storm arrived sooner than expected, the last place a person wanted to meet it was out on the Cortes Bank.

Certainly things had looked unpromising on the night of January 4. "It was blowing thirty-five knots out of the south," Long recalled. "I was getting up every half hour and hearing branches being torn off trees." "Even in the morning," Gerlach said, "we didn't know what we were actually doing. There was lightning. I was thinking, 'There's no way.'" While Collins helped the men plot the timing they'd need to pull off the trip, even he had been skeptical. "It was really a tight little schedule when all this could've come together," he said. "There were only a few surfable hours."

In the end, Long said, "we thought, 'Hey, we've gotta try. If we can pull it off these will be the biggest waves we've ever surfed in our entire lives.'" So at dawn on January 5 they left Dana Point Harbor in photographer Rob Brown's boat, a thirty-six-foot power catamaran customized for shooting in rough ocean conditions, with a mount to carry a Jet Ski. From the start, the trip was a battle. There was too much chop and surge to tow the second Jet Ski, so they had to ride it, trading off in the frigid, stormy water, all of them fighting seasickness. "The sea was a raging mess," Long said, describing his first shift. "Driving rain and squalls."

When they realized they had gone only fourteen miles in the first hour and a half, they knew they needed to pick it up. If they couldn't travel at twenty-five knots per hour, they would never outrun the storm—and there was a chance they'd arrive at Cortes too late in the day to ride a single wave. "We realized, 'We have to put our heads down and do this,'" Long said. "Outside San Clemente Island things finally started to calm down."

As they approached Cortes Bank they could see huge plumes of whitewater shooting up in the air, more than five miles in the distance. "When you can see it from that far away, you know it's big," Long said. At one p.m. they arrived.

"We were humbled fast," Parsons said, wincing, the vision of what greeted them still fresh in his mind. Cortes Bank was a swirling, furious expanse, a riot of water, a coliseum of giant waves. But they were right in the window of time they had hoped for, suspended between storms as a minus tide crept in, making the waves even more powerful. "All the elements were there," he continued. "But there was a super-raw component because the storm was so close. There was no easy ride." He paused and drank some of his ice tea. "The water just gave off this angry feeling, like '*Don't* mess with me right now.' Waves were hitting together and jacking up really high and bouncy and turbulent, and we all looked at each other like, 'Uh, we'd better reevaluate this.'" He laughed.

When I'd spoken to Long over the phone, he too stressed how vulnerable they had felt, alone in the middle of an aquatic maelstrom. "We spent the first hour just looking at it because it was still really messed up from the wind the night before. We had to actually get up the nerve to surf. And those first waves we rode were the biggest, windiest, scariest things ever. It was one of those days when you just could not make a mistake. Everything was on the line."

They were also uneasy in the knowledge that due to the last-minute

decision to go and the beat-the-clock urgency of the trip, they were woefully underequipped for emergencies. "I would've liked to have had extra
boats and extra Skis. It wasn't what you'd call a smart expedition," Long
said. "It was pretty rudimentary." As at least one added precaution the
men had doubled up on their flotation vests. "You're a bobbing top," Parsons said, chuckling. "And you need to be. The forces that are pressing
you down . . ."

In the event of a fall they had to hope for the best, and thankfully
that's what they got. "I had a pretty good wipeout," Parsons said, "and I
probably went, I don't know, half a football field underwater. Just rolling
and tumbling. It wasn't the worst pounding I've taken—I wasn't that
deep—but it was the most interesting because I went so far." When Parsons emerged from the rinse cycle, he was relieved to find Gerlach right
there to scoop him up. "But what if he hadn't seen where I fell or the
direction I went?" Parsons said with a frown. "I lost a surfboard on my
first trip to Cortes and it was just *gone*. I never saw that board again."

In the realm of accidents, he added, another potential nightmare was
to get hit by Cortes's lip: that was how spines and femurs got snapped.
"The worst thing is if it lands directly on you," Parsons emphasized. "You
want to be inside it, getting barreled, or out in front of it and get blasted."
As the lip came roaring from behind, he continued, a rider's every survival
mechanism went on red alert. "Your senses tell you where it is. I guess
noise plays a role but it's more feeling. You know the second it's gonna hit.
It's all timing."

On the boat, Rob Brown struggled to maneuver in the current without drifting into any impact zones, and to maintain visual contact with the
men while wrangling his camera gear and documenting the rides. He shot
stills; video was out of the question. "There were weird, rogue sets coming in," Brown said. "You had to know exactly where to sit." The vessel
dropped into troughs and pitched on swells and tossed in the whitewater,
and nothing about any part of the process was easy, but he managed to
photograph Parsons on an enormous face, his tiny figure silhouetted

against a hulking mountain of surf. A thick lip with an odd kink in it folded above him.

By all accounts, Long managed to catch a similar monster, but the ride had gone unrecorded. "I just felt the thing growing behind me," Long said. "You're not making any progress because the plane just keeps growing. You just have to point it straight." He knew he was running for his life because he could feel the lip behind him, closing in. "It broke and I was completely engulfed in whitewater. When I came out the other end my heart was in my throat."

The session ended with bodies, surfboards, and luck intact, and as darkness engulfed the bank they turned east for the long drive home. Looking back while a trace of light remained, they could see the next storm front skulking on the horizon, a menacing lead gray wall. A south wind swept across the water; rain began to fall. "You could see on every-one's faces," Long said. "We'd just pulled off the impossible." Though the men should have been exhausted to the point of collapse, adrenaline kept them alert. Wearing headlamps for vision, once again they took turns piloting the second Jet Ski home through the battering surf. "We got *pounded*," Parsons said, shaking his head. "The bumps come heavy at night when you can't see them. Every once in a while your face hits the handlebars. Chasing the boat I was thinking, 'Man, what if I hit a whale at this speed? Or what if I just fly off? There's no way they would see me.'" Then he brightened. "But you just feel so alive. So it was fun."

Greg Long stood at the podium as the cheers grew louder. The five finalists for Biggest Wave had been announced, and now it was only a matter of opening the envelope. Because he had won the award the pre-vious year for a sixty-foot wave at a place called Dungeons in South Africa, Long was doing the honors. When he looked down at the name of the winner, there was no trace of surprise on his face: "Mike Parsons—January 5, Cortes Bank!" Brown's photograph, somewhat grainy but still astonishing, splashed onto the screens and Parsons stood up, kissed his eight-months-pregnant wife, Tara, clapped Gerlach on the back, and headed for the stage. He stood there for a moment looking happy in his understated way, neat in a brown short-sleeved dress shirt and khakis.

"Wow, incredible," he said. "I'm really honored. That day was special. It was a bit of a mission." He thanked Billabong and Sean Collins and his family; he thanked Brown, and Gerlach—"My partner in crime for ten years now"—and he thanked Larry Moore, to whom he dedicated the award, for discovering the Cortes wave in the first place. "It's amazing how far we've come in ten years," he said. "And we're just getting started at this game. We're just getting the hang of it now." A cardboard check for $15,000 appeared at his side; someone wheeled a new Honda Jet Ski onto the stage. Then Bill Sharp came out and informed the audience that Parsons's wave was officially deemed to be seventy-feet-plus, "which would make it a new Guinness World Record!"

As the applause died down, Occy reappeared at the microphone. Festivities had been under way for several hours, and his voice had begun to slur. He looked up at the Cortes wave on the screen and turned to the audience. "These guys risk their lives to ride big waves," he said, with a nod of respect. "And that is jus' what they do."

Blinded by the glare of the spotlights, blurred by booze, and deafened by the din of the screaming logos, it was easy to lose sight of the real deal: the moment between the rider and the wave. Watching Greg Long up onstage applauding Parsons's world-record ride, I remembered something he'd said once, about what that moment was like. "You're just caught up in those few seconds and nothing else matters," Long had told me. "Sound, smell, everything just totally goes out the window. It's what's directly in front of you, what you need to do to make that wave, and *nothing* else." In giant waves, he'd added, "you're dealing with energies that are so much greater than you, or anything you've ever dealt with." That shared experience was what brought this group together once a year, to celebrate the rush of that moment—having had it, and having survived it. That was the bond, and yes, it was worth the next morning's hangover.

Standing on the sidelines, two of the event hostesses, charged with toting the blown-up cardboard checks for the winners onto the stage and adding a dash of sex appeal to the proceedings, watched the celebrating men. "I think the idea of surfing a one-hundred-foot wave is insane," said one, a blonde in a black miniskirt. "It is insane that they want to do it. They are insane."

"Yeah," said the other, a brunette, smiling slowly and shaking her hair. "But I've been doing this event now for three years, and each year the event gets bigger—and so do the waves."

As we watched the mountainous waves and felt them battering the ship and saw the impact of the gale-force wind, we realized the great strength of the elements— so strong and cruel they are and what little hope we humans have when we are in them.

Merlyn Wright,
passenger aboard the doomed ship Oceanos

THE WILD COAST

CAPE TOWN, SOUTH AFRICA

When a 120-foot rogue wave rears up in front of a 300,000-ton oil tanker, sucking the ship down into a black hole as it explodes on the bow, crushing the hydraulics and snapping the rudder so that steering is no longer an option, knocking out the engine and crumpling the deck like a tin can while gale winds drag it across the unforgiving shallows—well, the crew will want to get Captain Nicholas Sloane on its radio, or someone very much like him. And they will want to do it fast.

Sloane, a marine salvage expert, is based in Cape Town, South Africa. In his line of work—saving foundering ships from disaster—this is an excellent place to be. Right out Sloane's back door lies the Transkei, or Wild Coast, a five-hundred-mile stretch of Indian Ocean that runs from Durban down to the Cape. The area's distinguishing feature is the Agulhas Current, a fast-flowing pulse of water that streams south from the tip of Madagascar. Like the Gulf Stream off the U.S. East Coast, the Agulhas is a mighty and treacherous western boundary current, more than ten degrees warmer than the surrounding seas. It ranges from sixty to one hundred miles wide, its five-knot average velocity tempting ships to speed their journeys by riding along in its core. This can be a smart tactic or (as many boats have discovered) a highly ill-advised one. It all comes down to the weather: under certain conditions the Agulhas Current can be counted on to kick up the weirdest, wildest, most destructive waves in the world.

A number of factors conspire to make this happen. The current runs along the edge of the continental shelf, the land that slopes gradually beneath the ocean before dropping off into the deep abyssal plain. In South Africa that shelf happens to be narrow as well as steeply banked, slashed with canyons, and booby-trapped with shoals and shifting underwater sand dunes—all of which create whorls and eddies and assorted pockets of trouble. The Agulhas contains so many odd vortices, one oceanographer told me, that if you were to dump red dye into the water and view the current from space, it would look less like a purposeful river and more like the spinning teacup ride at Disneyland.

Cape Agulhas, the continent's southernmost point, has the added nastiness of two oceans colliding: the Indian and the South Atlantic. Fifteenth-century Portuguese explorers referred to this area as the Grave-yard of Ships, and the name stuck for good reason. It is here that the hot-blooded Agulhas Current smacks directly into the cold, dense southern swells that have stomped north from Antarctica, fueled by relentless winds. This oceanic clash of the titans creates enormous waves that are angry, unstable, and steep; a rogue-wave factory in one of the world's busiest shipping lanes. I had seen the pictures: tankers with their bows bashed in and their hulls punched open, as if by a massive fist. And those were the lucky ones. Hundreds more simply disappeared. Ships need rescuing from these waves constantly, and as a result many highly experienced marine salvagers hail from South Africa. For Sloane and his company, Svitzer, there is never any shortage of work.

I met Sloane in his office near the Cape Town Harbor one morning, catching him in the building's lobby on his way back from a court appearance. Even in a suit and tie he looked rugged, as though carved from some kind of igneous rock. Sloane had dark blond hair, pale hazel eyes, and a dry sense of humor. As we stepped onto the elevator he explained that marine salvage companies spent as much time in courtrooms as they did in the waves. Legal wrangling was standard, he said, because ship rescue was such a costly affair. Once a ship's Mayday had been broadcast, the ensuing salvage operation required serious up-front expense; the shuttling of heavy machinery and manpower was almost military in scope. They needed helicopters and tugboats, along with dredges, oil-containment booms, forklifts, pumps, hoses, cables, Zodiacs, Jet Skis, and decompression chambers. The salvage crew included hazmat specialists, chemists, pilots, mariners, ecologists, engineers, mechanics, meteorologists, wave forecasters, fire-risk experts, welders, medics, and immersion divers, among others.

Before a salvage company would even consider a job, the imperiled vessel's captain had to submit to an insurance agreement known as Lloyd's Open Form. Basically, the form stated that the salvager, having rescued the ship, was entitled to a portion of its value. The only question was the size of the claim. To determine this, facts were hashed out in the aftermath: How desperate had the situation been? How big were the waves? How likely was the ship to have gone down? How many lives were in jeopardy? How many gallons of oil had been safely corralled? Months and even years of tussling, in and out of courtrooms, could pass before an agreement was reached.

"When the boat's breaking up and the oil's headed for the beach and the authorities are going berserk, they really want you to come immediately," Sloane said. "But then a year later they start saying, 'Well, hold on, what's this invoice for?' Sometimes we don't get paid at all."

We stepped off the elevator onto a busy floor filled with men in suits who were working the phones from a warren of glass-walled offices and cubicles. The place was hopping, Sloane told me, because a tanker had just run aground off Mozambique and Svitzer was trying to win the job of safely dislodging it. There were four major salvage companies in the region and they would all be vying for the work, racing to round up crews and send them to the scene to fix the situation before the waves pounded the vessel to pieces, spilling 3,500 tons of diesel fuel near the mouth of the Zambezi River.

As we walked across the floor I heard bits of urgent conversation being murmured and barked into headsets:

"They want sixty thousand Namibian dollars to get her out," one man said, a frown etched on his brow and his shirtsleeves rolled up.

"Are you available to go up to Mozambique for some good problems?" another broker asked, hunched over a map.

"The tug's in Durban," a thin, frazzled-looking guy said, scrolling through files on his computer screen. "They're leaving now."

Sloane's office, in the corner, featured a poster-size photograph of a small figure in a survival suit rappelling down a two-hundred-foot wire that dangled from a helicopter. The figure's destination, near the bottom of the frame, was a bulk carrier that was tilted forty degrees to port, engulfed in boiling whitewater from the enormous waves that were flooding its deck. "Twenty-four-meter seas that day," Sloane said offhandedly, noticing me gaping at the picture.

"Is that *you*?"

"Yeah. We've had some fun."

Sloane had been in the business since 1984, working his way from the merchant marines to become a captain and then a salvage master and now a managing director in a global company, overseeing rapid-response crews in the hairiest situations. He'd been stationed on South Africa's brawny salvage tugs the *Wolraad Woltemade* and the *John Ross*, and he'd gotten a big eyeful of what the Agulhas Current could dish out in a storm. As I looked around his office at more framed photos of tankers exploding spectacularly or wallowing helplessly in monstrous seas, it was clear that Sloane's idea of "fun" involved a degree of mayhem that others would find less alluring. In this, he and the tow surfers were kindred spirits. The crazier the waves, the more eager they were to get out there. But a marine salvager's job came with even more hazards. On top of facing the ocean's wrath, he encountered all kinds of colorful man-made dangers.

When waves or fire or an unplanned encounter with rocks disabled a ship, the first question anyone asked was: What is the cargo? In happy scenarios the ship carried something that could spill harmlessly into the ocean, like wheat or frozen seafood, something inflammable or at least nontoxic. The most precarious situations arose when a damaged ship contained lethal, explosive chemicals like ammonia, toluene, or phenol (a common ingredient in plastics that can cause paralysis when its fumes are inhaled), to cite just a few. In cases like these the salvagers weighed their own safety against the knowledge that if they couldn't save the day, ten thousand tons of fungicide or acetone would soon be sluicing across the reef.

Earlier that morning I had read in the *Cape Times* newspaper about a salvage team that was in the Philippines trying to prevent the *Princess of the Stars*—a ferry that had capsized in a typhoon, killing more than eight hundred passengers—from leaking its illicit cargo of pesticides onto the shores of Sibuyan Island, a place so environmentally pristine that it was referred to as "the Galapagos of Asia." The chemicals, which shouldn't have been within miles of a passenger ferry, had been destined for a pineapple plantation. Until the salvagers stumbled across the ten-ton cache of endosulfan, an acutely poisonous pesticide that has been banned in over fifty countries, they hadn't known it was on board. The crew was dead; the ship's owners denied knowledge of its presence. All salvage work had been temporarily halted as the team planned how to extract the chemicals safely, racing the clock before the ship broke apart in heavy seas.

When I mentioned the *Princess of the Stars* to Sloane, he nodded soberly. Scenes like this were typical, he said. The most lethal substances were subject to such tight restrictions that some shippers simply didn't declare them: "They try to hide away the really dangerous stuff." The more threatening the substance and the slipperier the exporter, therefore, the more likely the salvagers wouldn't realize what they were dealing with until they were staring at it. Even then, they couldn't be sure. Sloane recalled one case in which a hold filled with cyanide powder had been labeled as flour.

Other chemicals, while not life-threatening, had what Sloane referred to as a high "nuisance factor" in a spill, meaning they smelled terrible or fouled the water temporarily, causing beaches to be closed. Then there was oil. Crude, diesel, jet fuel, liquefied natural gas: oil in all its forms was a heartbreaking, infuriating, and all-too-common sight in the ocean. Supertankers, behemoths that couldn't make it through the Suez Canal, swung down from the Middle East, took their chances hopping a ride in the Agulhas, and met their share of disasters. Salvagers used every tool at their disposal to prevent the damaged tankers from gushing out their contents, especially in fragile near-shore environments, but sometimes the battle was lost. Sloane wheeled around in his chair and gestured to another picture of a wrecked ship, its rusty bow jutting from the water at a sharp angle. "That was right off the coast here," he said. "We had fourteen thousand penguins that we had to catch and wash."

Compared to those kinds of horrors, dealing with giant waves must have seemed downright pleasant, though no less threatening. Sloane recalled looking uneasily out of a helicopter that was hovering about one hundred feet above a disabled ship as eighty-foot sets broncoed through, worrying about a 110-foot rogue rearing up from behind, or seawater dousing the turbines and knocking out the engine. Meanwhile, down on the water, wrist-thick metal cables whipped through the air and heavy machinery jostled around in stormy seas. Hands, fingers, toes, eyes: they could be lost with ease. "Everyone's been hurt at some stage," Sloane said. "You always end up with a few broken bones. When we're out there we're pushing the limits because the people who are normally crew on the ship, they've lost control of the situation. And that's where we come in: when it's already got to a point where it's really dangerous."

Ships have been meeting giant waves in the Agulhas Current as far back as you care to page in the history books. Often the boats perished, but those that escaped brought back strikingly similar stories: Gale winds had been blowing, occasionally switching direction. Conditions were seriously lumpy, but every so often a far larger wave (or a set of them) would lurch from the sea. Uncharted eddies dragged the vessels off course and deep troughs yawned in front of them and waves came from every direction. If a ship found itself along the three-hundred-mile stretch between East London and Port Elizabeth in heavy weather, all bets were off. If the Agulhas Current could be counted on for a steady supply of freak waves, this section was the most reliable of all.

Consider this account of the *São João*, a Portuguese galleon that was heading home to Lisbon in 1552. Near Port Elizabeth, a storm hit: "The pilot, Andre Vas, was steering a course for Cape Agulhas, which they duly sighted, but then they encountered easterly gales which blew them to about 65 nautical miles southwest of the Cape of Good Hope . . . Next, beset by furious westerly gales, the captain, master, and pilot agreed that it would be best to run before the storm, back in an easterly direction . . . About 340 miles east of the Cape the wind shifted to the east again and they resumed their westward voyage in a heavy swell that threatened to sink the galleon at any moment . . . Unfortunately another westerly storm unleashed its fury upon them, the ship veered around and three huge waves struck her abeam, breaking all the shrouds and backstays on the starboard side. It was now decided to cut away the mainmast, but while this was being done, it snapped . . . and everything disappeared over the side. Now the rudder broke in half and was carried away." The *São João* didn't make it (though some passengers struggled to shore); nor did the thousand other ships that fill the pages of *Shipwrecks and Salvage in South Africa*, the reference book from which the preceding passage was taken. These were merciless waters.

One of the most unnerving incidents in the Agulhas involved a ship called the *Waratah*, which left port from Durban on the evening of

July 26, 1909. Its intended journey would take it to Cape Town and then on to England. Now known as the *Titanic* of the South, the five-hundred-foot-long, 9,300-ton ship, designed to carry both passengers and cargo on the long-haul trip from Britain to Australia, had been launched in 1908 and rated 100 A1 by Lloyd's of London, the most enthusiastic thumbs-up available. At *Waratah*'s helm was Captain Josiah Edward Ilbery, sixty-nine, a distinguished seaman who had attained the rank of commodore. Even gazing out from old photographs, Ilbery inspires confidence. He was silver haired and extravagantly bewhiskered, with piercingly clear eyes and a strong chin; precisely the type of heroic old salt you'd want steering your boat through the Agulhas Current. Ilbery, however, made at least one very big mistake.

In the pre-satellite, pre-Surfline, pre-GPS, pre–emergency beacon, pre-radio days of old, captains used whatever scarce information was available to make their weather calls. Likely, Ilbery was unaware of the worsening conditions when he left Durban to steam down the Wild Coast. It wouldn't have been long, though, before he realized the truth: other ships in the area had been bombarded by such high seas that they'd dumped their cargos. The *Waratah*, along with its 211 passengers and crew, was carrying 6,650 tons of supplies that included a fresh store of coal and, not very promisingly, 1,300 tons of lead.

(Actually, on that night there were supposed to be 212 people aboard the ship. One man, an engineer named Claude Sawyer, had disembarked at Durban, refusing to go any farther. Sawyer, who unsuccessfully tried to convince other passengers to get off the *Waratah* along with him, was outspoken about "the strange way the ship dealt with the waves" on the passage back from Australia. To make matters worse, Sawyer had been plagued by a ghostly vision: an angry man with long matted hair, emerging from the sea waving a bloody sword and calling *"Waratah, Waratah."* Sawyer was not shy about his misgivings and by all accounts spent much of the voyage spreading them. He complained at dinner, during the lifeboat drill, and in telegrams to his wife back in England. As the ship steamed away from Durban, the others must have felt glad to be rid of him.)

The *Waratah* was last seen at six a.m. on July 27, when she overtook a smaller ship just north of East London. Both vessels were battling thirty- and forty-foot waves, and they acknowledged each other as they passed. Then the *Waratah* sailed into oblivion. Months of extensive searching by the Australian, South African, and British navies, along with salvage crews and other ships, turned up nothing, not even a scrap of wreckage. Several ships reported seeing bodies floating in the water, including such specific visions as a little blond girl in a red dress, but all were discredited. A *Waratah* life ring washed up in New Zealand, though it could have fallen overboard at any time. More unsettling, a week after the ship's disappearance a bedraggled, confused man was found wandering on a South African beach; the only words he would say were *"Waratah"* and "big wave." When no elaboration came forth, he was packed off to a mental hospital, his tale of survival (if in fact he had one) remaining locked in his mind. It wasn't until December 15, 1909, almost five months after the *Waratah* vanished, that the searches were finally halted and the Lutine Bell at Lloyd's of London sounded its mournful call.

In recent years the *Waratah*, like the *Titanic*, has attracted its share of seekers. A number of attempts were made to locate its wreckage (one bankrolled by the American novelist and underwater explorer Clive Cussler). It looked as though these efforts would be rewarded in 1999, when side-scan sonar traced the outline of a sand-covered ship that matched the *Waratah*'s contours. The sunken vessel lay four miles offshore in 580 feet of water just north of East London, close to where it had last been seen.

Champagne was cracked, a press release issued. A submarine dived to film the ship in its final resting place for an eventual feature film. Only problem was, when the wreck loomed into closer view it wasn't the *Waratah* at all, but rather a transport ship of World War II vintage with a deck full of tanks and rubber tires. Similarly, another wreck lying nearby was not the *Waratah* either. This one was the cruise ship *Oceanos*, which went down in a storm on August 4, 1991.

In a spectacular air and sea rescue by the South African Air Force and Navy, all 571 of the *Oceanos*'s passengers and crew had been evacuated after a giant wave breached the hull, flooding the engine room and knocking out power. Scandal erupted immediately after the incident due to the fact that the captain, Yiannis Avranas, fifty-one, had elbowed aside children, women, and the elderly (including an eighty-year-old woman with a broken hip) to ensure himself a seat on the first rescue helicopter. Similarly, the rest of Avranas's senior crew distinguished themselves by commandeering the sturdiest lifeboat and bailing out, luggage and all, before most passengers were even aware that the ship was sinking. With the captain and crew gone, the *Oceanos* filling up with water, and hundreds of people still on board, a fast-thinking cruise director radioed an SOS. It fell to the ship's comedian, magician, and lounge band musicians to supervise the rescue, a terrifying high-wire act conducted in mountainous swells and fifty-knot winds.

It was miraculous that no one died. Later charged with negligence, Avranas protested that he had done nothing wrong. "When I order abandon ship, it doesn't matter what time I leave," he told ABC News angrily. "Abandon is for everybody. If some people like to stay, they can stay."

I'd heard about the *Oceanos* from Sloane, who'd been in the Agulhas himself that night, removing four hundred people from an oil platform that was in danger of being ripped from its moorings. These were the most out-of-control conditions anyone could remember seeing on the Wild Coast, with three major salvage operations under way at the same time. "Eighty-five knots of wind," Sloane recalled, grimacing. "The *average* wave height that night was twenty-three, twenty-four meters [75–80 feet]." He recalled watching with alarm as the extending gangway they were using to get people off the rig, one hundred feet above the water, barely escaped being swept away by a wave. "They were really close to capsizing," Sloane continued. "While we were out there the call came in from the *Oceanos*. And then the *Mimosa*—another ship, an oil tanker—got in trouble. That was the biggest storm I've ever been in."

"Have you ever seen a wave that really terrified you?" I asked. "Something totally off the charts?"

Sloane nodded. "Oh, yeah. That night I did." He pulled up a picture on his computer and turned the screen to face me. It showed a supertanker almost completely submerged in a haze of spray, being engulfed by waves that were washing over its deck—at least seventy feet above the waterline. "They said the one-hundred-foot wave would never happen," Sloane said, smirking slightly. "Well, they were wrong."

At Sloane's suggestion I had met with Captain Dai Davies, a renowned salvage master who had seen more of the Agulhas's extreme waves than anyone. Davies, a trim man in his late seventies, still exuded the air of total competence that had distinguished his long career. He seemed to have a photographic memory for names, dates, ships, and storms, instantly recalling such arcane details as the nationality of a tanker's crew, the type of cargo it was carrying, and what had been served for dinner during the rescue. The events of August 4, 1991, were crystal clear in his mind.

"The *Mimosa*," he said, "she was three-hundred-sixty-five-thousand tons. Norwegian crew. The ship was in trouble off the other side of Port Elizabeth from here, and she was coming down the coast. A big wave struck her. And I got the call." He shook his head. "I'll never forget it. The weather was horrific, horrific, horrific!" Speaking to the captain by radio, Davies learned the wave had destroyed the tanker's hydraulics, jamming its rudder and disabling its steering. "The captain said this wave was *so* big. He saw it from the bridge. The waves at the time were sixty feet, and this one was twice as big again. Just came out of nowhere. Extra-deep trough. They went down into that."

With great difficulty, the ship rolling wildly in the maelstrom of waves, swells, and current, Davies and his men successfully lassoed the *Mimosa* (with its thousands of tons of oil) and towed it into sheltered waters, then eventually all the way to Dubai, where it was repaired. "Not a drop of oil was spilled," Davies said proudly. "Not a drop."

He went on to recite a list of ships that had met up with extreme waves in these waters, describing the damage to each in vivid terms: "It looked as if a cutting torch had split the ship in half entirely!" Then: "The ship's side plating was punched in, completely smashed in, causing a hole to be formed you could fit three double-decker buses into." Finally: "I looked down and I could see that the bow was gone! Four and a half tons of steel! It had dropped off."

"We've got a very queer current situation on this coast," Davies said, and then paused for effect. After a moment, he continued, his brisk Welsh brogue dialed lower and raspier. "I call this part of the ocean the Final Surveyor. If ships can get past here, they're okay, you understand? But a lot of them don't. A lot of ships get beaten to a pulp."

On a slate gray day I left Cape Town and drove south on a winding highway, past the hamlets of Muizenberg and Kalk Bay and Fish Hoek, until I got to Simon's Town, a pretty community perched at the edge of False Bay, only five miles from the Cape of Good Hope. There I took a hard right and switchbacked my way up into Table Mountain National Park. It was lonely at the top of the plateau, scrubby and windswept, the quaint houses and restaurants having been left far below. "Nuisance Grave," a sign announced, unappealingly. As I drove, it became clear that the trees and vegetation weren't merely scruffy; the landscape had been charred black from fire. This was a narrow peninsula, pinching off with dramatic finality at the cape, and I could see a glint of ocean both in front of me and behind.

Cresting the rise, I passed a sprawling shantytown set among the burned-out trees; a tarpaper, plywood, and tin encampment in what seemed like the middle of nowhere. I doubled back, realizing I'd overshot my turnoff, and this time I found what I was looking for: a large metal gate that opened onto a dirt driveway. Jean Pierre Arabonis had seen my car go by the first time and he stood out front, waiting. Opening the gate, he waved me in. I rolled up the driveway and stopped in front of his office, a smallish, low-ceilinged structure made of buff-colored stone. A fifty-foot tower stood next to it, the steel latticework climbing toward a wooden platform that held an enormous satellite receiving dish, pointed toward the heavens.

Here, in this unlikely place, was the headquarters of Arabonis's company, Ocean Satellite Imaging Systems (OSIS). A compact, thirty-seven-year-old South African of Belgian descent, Arabonis was a maritime meteorologist who was in great demand for his eerily accurate wave, ocean, and weather forecasts. He had clients all over the world—fishing fleets, shipping companies, government agencies, marine salvagers—but he specialized in South Africa's complex waters. When other forecasters were zigging, Arabonis often zagged, and when he was proven right time after time, his reputation grew.

We shook hands and he showed me into the office, where another meteorologist, Mark Stonestreet, sat at a computer, a nautical chart unfurled beside him. I had come to meet Arabonis because I'd heard that he understood the Agulhas Current's freak waves better than anyone; that, in fact, he released warnings when they were likely to appear so that ships could rethink their routes. Once in 1995 Arabonis had forecast one-hundred-foot waves in the Transkei (near East London) with such precision that Sloane had been able to fly a tow surfer named Jason Ribbink into the Agulhas by helicopter, then winch him, his driver, and their Jet Ski into exact position to surf the current at peak fury.

"I was busy with this job," Sloane told me, recounting the incident. "We had the *Kiperousa*, a Greek bulk carrier, grounded on the beach. Heavy surf; she was in danger of breaking up. So I was keeping in touch with Jean Pierre about what the weather was doing. He said: 'There are going to be some really abnormal waves offshore.' Well, then about an hour later Jason called to say he was planning to surf the big waves at Dungeons, you know, near Hout Bay. I said, 'Well, you shouldn't be *there*—you should be up the Transkei coast! JP says it's gonna be a hundred feet.'"

Ribbink and his tow partner, Dane Patterson, took him up on the tip, driving north. They met up with Sloane, and in East London a safety diver joined the group. Curious more than anything, Sloane strapped their Jet Ski beneath the helicopter and lit out over the Indian Ocean with the entire crew. "We flew past a couple ships," Sloane recalled. "There's nothing out there, and here we are with a Jet Ski dangling. They called us up and said, 'Where the hell are you guys going?' I said, 'Oh, just heading out for a little surfing!'" He laughed. "Basically, what we did that day was break all these aviation rules. I was glad it never hit the press."

Thirty miles offshore they arrived at Arabonis's appointed place, an area where the storm energy met a forceful swirl in the current. "We released the Jet Ski from the helicopter and the driver jumped into the water," Sloane said. "When he was ready, Jason jumped out with his surfboard." In that specific spot, Arabonis had told them, "you'll have a two-hour window for one-hundred-foot waves."

He was right.

"Oh yeah, they got some," Sloane confirmed, raising his eyebrows for emphasis. Unfortunately, viewed from the air, the waves didn't look nearly as awe-inspiring as they actually were; photographs couldn't capture the spectacle. In the open ocean a swell doesn't break the way it does on a reef or a seamount. You don't get the fearsome curling lip or the horrifying hang time as the wave winds up to release its energy in one knockout punch. Offshore in the Agulhas the waves bucked and rolled and tossed their weight around all right, at enormous heights and with terrible power, but they resembled endless ramps more than steep cliffs. "These waves have a huge long wavelength so they don't look like a heck of a lot," Arabonis said. "And you can't see them as well when the whole area is rough—on that day a forty- or fifty-knot wind was blowing."

Thumbing through a map drawer for a chart of the current, Arabonis kept talking. "I don't work by hard-and-fast rules," he explained. "I

work pretty much on gut feel, and on what I've seen happen before."
When his inner one-hundred-foot-wave alarm bells sounded, two simple
things had usually set them off: "Southwesterly swell [at an interval of]
more than fourteen seconds, and seas more than fifteen feet high." If
those two variables were present, then the ruckus could potentially begin.
But the abnormal waves themselves, should they appear, defied any kind
of neat explanation.

"This is where all the wave mechanics start becoming a little fuzzy,"
Arabonis said. A perplexed look crossed his face at the impossibility of
knowing every last wave secret. "These freaks . . . ," he said, drawing out
the words and then beginning the thought anew. "Well, it's not oceanographers looking at them anymore. It's physicists! Because they've discovered that these waves are behaving in a manner that is similar to light
waves. They can suck the energy from both sides and concentrate it in
one spot. And light waves are partially particles and partially wavelike.
It's moving [the study of waves] into a whole different dimension."

In this alternate universe of ocean behavior, individual waves in the
Agulhas somehow reached a tipping point where (once Arabonis's two
basic requirements had been met) a third (usually unknown) element
entered the picture and upended the whole equation. Suddenly things
went nonlinear. They went off the charts, off the radar, off the neatly
plotted curves of statistical wave-height distribution, and into the shadowy, destructive territory that the Agulhas Current shared with the
Bermuda Triangle, where things—enormous things like supertankers
and awfully big things like cruise ships and smaller-but-still-hard-to-
misplace things like eighty-foot yachts—disappeared into the maw.

I told Arabonis I'd read the statistic that, on average, two large ships
a week go missing in the global seas. "The figure I heard was that one bulk
carrier a week was disappearing," he replied, with a rush of explanation.
"Iron-ore carriers. Those things are death traps. They're built to poor
specification, a lot of them are quite old, and they lie very low in the water.
The waves generally break the first and second hatch covers. Once you

breach those, two things can happen: the bulkheads collapse, or the vessel starts taking a nose-down attitude. It just floods from stem to stern and powers itself to the bottom. It can sink in approximately one minute."

As he outlined this harsh fate he turned to a blackboard and brusquely diagrammed the scenario, his chalk snapping as he sketched the outlines of the doomed bulker. He stepped back to examine his drawing. "We'll likely see more incidents like this as the price of commodities—and the demand—remains high. The older ships are remaining in service. Normally a bulk carrier, after twenty years she needs to be broken up. But now there are many rust buckets still plying their trade after twenty-five, twenty-seven years. They get a third-world crew and a captain who's two warnings away from losing his ticket. They should have been off the seas long ago."

A strong wind gusted past, rattling the windows. It was a raw day to start, but it felt even more so when considered from up here, at what seemed like the end of the world. Five miles south, the cape jutted into the South Atlantic; after that, next stop: Antarctica. But Arabonis didn't mind the isolation. In fact, he needed it for his satellite reception. He did mind living next door to a squatter camp, though, and over the years he had defended his home, his family, his office, his computers, his electronic equipment, and his dog from armed attacks, attempted home invasions, and glue-sniffing, knife-waving intruders. The scorched landscape, he told me, was the result of arson.

But the same things that had drawn Arabonis to his profession made South Africa an ideal home. Along with his marine-based career, he led a richly aquatic life. He was a licensed yacht master, an experienced sailor, a Class IV commercial diver, and a scuba instructor with more than two thousand dives under his belt. To hear Arabonis describe his undersea excursions is to learn that these waters did not lack for sharks. For instance, at nearby Pyramid Rock, only three hundred yards off a beach near Simon's Town, Arabonis regularly encountered clusters of them. "I've seen three great whites," he said. "All three were quite scary." He also ran into spotted seven-gill cow sharks, a creature whose mellow-sounding name belied its aggressive personality. "These things are not to be fooled with," he cautioned, as though I were planning to head down there right now for a dip.

Only a few weeks before my visit, all of Arabonis's interests had collided in a single, bizarre tragedy, an accident involving a giant wave, a boat, and a group that was diving with great white sharks, a stone's throw up the coast at Dyer Island. The thirty-five-foot catamaran, owned by the cage-diving operator Shark Team, had motored out on the fateful morning carrying ten divers and a nine-person crew. Conditions bordered on crummy, with a seven-foot southwest swell buffeted by a ten-knot southeast wind under cement-colored skies. Wads of kelp floated on the surface, borne on the swell. As they dropped anchor in an area known as Shark Alley, the chumming began, and some great whites had already been sighted circling at ten-fifteen when, as one diver described it later, "this huge wave blacked out the sky."

The wave picked the catamaran up like a piece of lint and tossed it (with its attached dive cages) upside down, throwing some people clear but pinning most beneath the hull. Several boats that had been anchored nearby raced over to pull the struggling divers from the sharky waters, but not everyone was saved. Three people were drowned, knocked unconscious by the impact or tangled in the boat's ropes; six more were seriously injured.

In the aftermath, and with so many eyewitnesses, one thing became certain: the wave had been at least three times the size of the surrounding seas. "I've never seen anything like that wave," an onlooker said. "Nothing would have stood a chance, except maybe an ocean liner." "Without a shadow of a doubt it was a freak or rogue wave," said a National Sea and Rescue Institute spokesman. "You have to be humble in the ocean. It's a place where the unknown happens."

To chalk up the accident to nature's mysteries, however, did not suffice for insurance companies, who in turn hired Arabonis to find out what exactly had happened out there. Arabonis hated forensic jobs—he was often called upon to help locate missing yachts, situations that usually came with unhappy endings—but he duly produced a forty-two-page report about the day: the sea state, weather, bathymetry, tides, water depth, even the moon phase. The wave, he concluded, had been about twenty feet high, and despite popular opinion, Arabonis believed there was nothing too murky about it. This was simply a once-in-a-blue-moon wave, a much bigger animal that shrieked out of the background at long intervals, rare but explainable ocean behavior. "Small boats get into trouble in big waves," he said.

So what causes a garden-variety six-foot wave to grow to such an extent that it can start flipping boats? After examining charts and photographs of how the seas were breaking that day, Arabonis concluded that a shoal near Shark Alley had focused the wave energy, the way a magnifying glass can amplify light energy if held in just the right position. "When the waves bunch up, then you're going to get the odd big one," he said. "But what is happening, I'm almost one hundred percent certain, is that every now and then, the bottom features are sufficient for it to bunch up *plus* focus at one point. That's your little tie-breaker. That's what pushes it over the edge so you get your one-in-a-thousand wave."

This same phenomenon could be scaled up and exported into the Agulhas, where waves could snap tankers in half. "I suspect the discontinuities in the [continental] shelf are causing—are assisting—these abnormal waves to form," Arabonis said. The underwater canyons and irregular slopes snagged the swells, slowing parts of them down and causing pileups that were whipped to even meaner heights by an opposing wind, and by the head butting that went on between the current and the oncoming South Atlantic swells. If you set out in a lab to create the ideal environment for mutant waves, you couldn't do any better. "The ocean's a pretty wild place," Arabonis said soberly. "A lot of ships do get lost. And if you consider the smaller boats, the numbers are extraordinary. I heard a statistic of several thousand yachts a year disappearing." He stood to refill our cups of tea. "A wave that's sixty feet from trough to crest is a very scary beast.

But when it hits seventy-five feet you're talking about something absolutely out of this world."

Both at the surface and below, it seems, everything is in constant flux. Energy flows and surges and occasionally bellows. Water itself is a most complex substance, eight hundred times denser than air, prone to confounding behavior. Wind is invisible but can wreak havoc wherever it goes. When it comes to making a wave, so many factors come into play that it is hard to know where any one thing ends and another begins, but in South Africa they measure their progress by the number of ships—and lives—they are able to save; the gallons of oil they prevent from hitting the drink, fouling the landscape. This is a full-time job.

"Knowing what they do about this place, why do ships come here at all?" I asked.

Arabonis let out a deep sigh, as though the question exasperated him. "The Suez Canal can handle only so many boats and only of a certain size," he said. "We're looking at thirty percent of the world's shipping going past the cape here." He swept his arm toward the window and beyond, to South Africa's Wild Coast. "They've got no other choice!"

The helicopter twisted off the ground at Cape Town International Airport, rising high above the Khayelitsha Township, a patchy sea of flat-roofed shacks and shanties that housed more than two million people, one of South Africa's starkest physical reminders of apartheid's cruel legacy. In the distance Table Mountain hunkered over the city, clouds swirling at its peak. We flew past Lion's Head, circled, and hovered in the silvery dusk light, then we swung out over the ocean, flying fast and low across the water. Waves streamed in, endless, and in front of us the African coastline wound its way toward the Indian Ocean, toward the grounded ship in Mozambique and the sunken remains of so many others.

I heard Sloane's voice in my headset. "Look down," he said, pointing to my window.

Below us, lying hard on the rocks of a promontory, was the battered skeleton of a ship. It was busted and listing heavily, with its midsection caved in. A dented crane hung crookedly off its deck. "A Russian ship was towing it down from the Congo," Sloane shouted, above the din of the rotors. "Lost it in a storm and it landed on the rocks. Fifty-foot seas. It was a hundred-million-dollar loss."

The helicopter dipped low so I could get a better look. The waves were hammering, and whitewater sprayed over the ship. I could see the patina of rust and decay, the proud insignia faded to shadow. When it was built the vessel had been stalwart and kingly; now it was lost to the elements. Sloane, Arabonis, Davies, and the other salvagers I'd spoken to here—they all expected a stormier future, more ships on the rocks. "The dynamics of the ocean are changing," Arabonis said. "There's more energy in all the systems." But looking down at the ruined freighter, I realized that one thing remained the same: the waves always won.

THAT WHICH IS, IS FAR OFF, AND DEEP, VERY DEEP.

WHO CAN FIND IT OUT?

Ecclesiastes 7:24

AT THE EDGE OF THE HORIZON

HAIKU, MAUI

Get your vest."

Hamilton's voice cut in and out over the cell phone, and I could hear a roaring noise in the background, as though he were speaking from inside a wind tunnel. Those three words brought things into quick focus: "vest" meant flotation vest, which in turn meant big waves. Given that we were on Maui, big waves meant Jaws. "Get your vest" therefore was shorthand for "Jaws is breaking." From the background sounds and Hamilton's brusque tone, I knew he was either calling from his truck, lead-footing it down the Hana Highway to launch from Ilima Kalama's house, or he was out on the water already, racing up the coast on a Jet Ski.

"Are you out there?" I asked, still foggy from sleep. It was just past dawn, and I could hear my neighbor's rooster screeching in the background. A wash of apricot light played over the Pacific. From my bedroom window everything looked deceptively peaceful. "How is it? How big is the——?"

He cut me off. "Just get down to Ilima's. But you gotta move it. Forget the mascara."

I jumped out of bed and pulled on a bathing suit, wet suit, and rashguard, grabbed my vest, and ran from the house still brushing my teeth, almost stepping on my cat. If Jaws was performing today, I wasn't going to miss it. Pulling out of my driveway, I noticed a handful of trucks and cars winding down Pe'ahi Road, a one-lane rut that snaked through ragged brush and neglected fields, dead-ending at a bluff above Jaws, two valleys away from Hamilton's cliffside vantage point. Word was out: the waves were here. Within hours a hundred people would be gathered at the edge of the lookout.

The swell, whatever size it was, came as a surprise to me. As recently as the previous day there had been no talk of exceptional waves showing up in the near future. There were no obvious magenta blobs advancing on Hawaii, no plans for Don Shearer and his helicopter to be on call. It wasn't unusual, however, for Hamilton to sense that conditions were changing—and say nothing about it. To his mind, making pronouncements about the ocean's future behavior was the ultimate arrogance. One of the fastest ways to aggravate him was to hold forth about what the waves would be doing next week, say, or later this season. "Tomorrow's going to be great," I made the mistake of saying one time, after looking at promising wave forecasts. "Oh, it *is*, is it?" he'd snapped in a sarcastic voice, turning on me with a hard look. "We don't *know* that. Nobody *knows* that." From this philosophy stemmed an aversion to making plans of any kind, a need for all options to remain open until the ocean actually showed its hand. "Forecasting's a crapshoot. I wait until I see the whites of the eyes," Hamilton said, describing how he judged what was, and was not, a worthy swell.

But something was up this morning. I could hear it in his voice. I rushed along the north shore to find out what it was.

Ilima's place was hidden from the road, accessed by a narrow, unmarked opening in a head-high field of sugarcane. Red mud caked my tires and my feet as I jumped out to unlatch the gate, then parked in the shade of a royal palm. Hamilton's black pickup was pulled off to the side, a trailer unhitched beside it. Several more trucks and Jet Skis dotted the lawn, which quickly merged into Baldwin Beach, a windward-facing, mile-long crescent of pale sand. Spreckelsville lay offshore just west of here; to the east Hookipa broke and a few miles beyond that, Jaws. Standing on the beach now, I saw that on the water things were nowhere near as tranquil as they'd looked from afar.

The ocean pitched and surged from the incoming swell, distinct layers of green and blue, light and dark, confused and choppy and frothy with whitewater. It was a bit of a haul to get from Ilima's to Pe'ahi by water, but launching from the beach was vital when there were waves: Jet Skis did not fare well when dropped into heavy seas next to rocky cliffs. Most tow teams put in at Maliko Gulch, a partially sheltered bay just up the coast, but even there the swell and backwash could make it impossible to pull a trailer up to the water. On the biggest days, Lickle had told me, one of the most daunting challenges at Jaws was getting out to the wave in the first place.

Not surprisingly, there was no sign of Hamilton, but down at the water's edge I saw a tall, lean figure wrestling a Jet Ski into the shallows. It was Don King, the Oahu-based cinematographer who had filmed most of Hamilton's movies. If you've ever watched stormy ocean footage shot from a fish-eye view in a major motion picture—the scary moments in *Castaway*, for instance, when Tom Hanks is flailing in heavy surf during a lightning squall—you've seen King's work. A former champion water polo player at Stanford, King could swim his camera into any kind of liquid mayhem and remain calmly in control. After having one of his photographs published on the cover of *Surfing Magazine* at age fifteen, King went on to pioneer the practice of popping up *inside* a wave to get unusual shots. The maneuver changed surf photography for good, though less adept swimmers attempted it at their peril.

I ran down to greet King and help him shove the two-thousand-pound machine off the beach. He was headed out, he said, strapping his camera case onto the rescue sled, and I could ride with him. "What's going on out there?" I asked. "Do you know?" King reached into the pocket of his surf shorts and pulled out his iPhone, which showed the latest readings from Buoy One, just north of Kauai. "Thirteen feet at nineteen seconds," he said with a nod. For all his daring, King was an understated guy. He had a compassionate face, his aquiline features framed by a mustache and glasses. Even in the most intense situations, King always had a gentle way of speaking and of carrying himself.

Pulling on my vest, I climbed onto the Ski. The wind had picked up a little, shivering the palm trees and blowing a fine spray back on us as King steered through the shore break. Immediately I felt the effects of the swell; triple overheads lurched in front of us, and farther out the entire ocean seemed to be heaving in giant, gulping breaths. "Hang on," he said, with a quick backward glance, and then he opened the throttle, racing straight at a menacing wave, cresting only a half second before it broke. I tightened my grip on the Ski.

"You've done this before, right?" King yelled, his voice muffled by the wind and surf. Before I could answer, he pulled a hairpin turn and ran back in the direction we'd come, a calculated retreat to escape another breaking wave. They were everywhere, the size of houses, rising from the left and right. As we wove through the patchwork reef, it was like we were running a gauntlet in some deranged ocean video game; the more successfully we dodged one wave, the quicker another sprang up.

King ricocheted through the shorebreak, darting forward and back, and I clenched every available muscle to make sure I stayed on the Ski. I'd heard stories from Mike Prickett, Sonny Miller, and others about being ejected in situations like this, vaulted through the air in a position that Lickle referred to as the "full Superman." Eventually King and I made it to deeper water, where the colors darkened from aquamarine and emerald to a fathomless navy, and the chop settled into a rolling swell. It was lumpy but, as yet, unthreatening. Making ten-knot progress up the coast, we motored on. Overhead, the pale sky filled with enough clouds to keep things interesting, puffy and regal with gray-tinged bellies.

"Have you seen Exploding Rock?" King asked, easing up and veering toward an area where a jagged lava formation created a dramatic surf break, spray geysering into the air. I knew the place—in fact I'd swum around in there, but under far calmer conditions. Today it was going off. A wave would sweep beneath the Jet Ski with the silky power of a baseball pitcher's perfect slider; when it connected with these rocks, all of its energy was blasted skyward. Sun glinted through the fifty-foot curtain of water, casting a scrim of tiny diamonds. At the edges, circular rainbows called glories shimmered like haloes. It was a spectacle as dreamlike as it was dangerous, an immense rush focused in one small spot.

We drove on. The ocean was alive with little peaks. No wonder the scientists were confused, I thought. This was anarchy. Every swell was born in a different place, made from a specific recipe of wind, time, and water, and as Hamilton had pointed out, each wave was unique as a fingerprint. It had its own provenance and its own destiny, clashing against its neighbors or merging with them, leaping out of the seascape or dissolving back into it.

When we rounded the last bay before Jaws, the sun hit my eyes and turned the people lining the cliff into toy-soldier silhouettes. So much water was moving, sucking the Jet Ski forward on a sloping trajectory toward the channel, and so much spray misted the air, that it was easy to become disoriented. My heart rate soared on adrenaline. I could hear the baritone booms, and I could smell the odd, faintly electrical scent that arose when water and storm energy met—but for some reason I couldn't see the wave. In the next second I realized why: King had come up on Jaws from behind. The roaring face was directly in front of us, but we were looking at its backside. The wave first emerged from the sea as an immense bulge—a perfectly rounded hillside that happened to be moving at about thirty miles per hour. When it tripped on the reef it sprang up and lunged open, detonating into whitewater when everything came smashing down. We were just outside the lineup where a dozen tow teams jockeyed for position, Jet Skis flashing in the glare. Standing as he piloted the Ski through the surf, King looked nonchalant, as though he were cruising through a parking lot.

Another Jet Ski approached us; Sierra Emory was driving, towing Hamilton. They didn't stop. As they passed I saw a look in Hamilton's eyes that I recognized, and it didn't allow for a midmorning chat. He was dressed in ninja black, no cheery yellow rash guard or flashy red flotation vest, and the muscles in his legs and back and forearms strained visibly against the rope, as though spoiling for battle. Realizing that Hamilton was about to ride a wave, King hightailed it back to the channel.

We pulled into the so-called safe zone, next to a pair of tow teams that had sidelined themselves, and as King reached past me to untether his camera case and assemble his housing, I stared at Jaws. At about forty feet, this wasn't the biggest day on record, but somehow that diminished nothing. The wave was breathtaking. As it rose, its face opened up to the cliffs and its lip curled over a full-bellied barrel. Except for luminous glints of turquoise at its peak, the wave was sapphire blue, gin clear, and flecked with white. If heaven were a color, it would be tinted like this. You could fall into this water and happily never come out and you could see it forever and never get tired of looking. Jaws did not permit its spectators to daydream about being someplace else, to feel bored or irritated or jaded. Watching it was an instant antidote to petty problems. There could be no confusion about who called the shots out here, at this gorgeous, haunted, heavy, lush, primordial place, with all its unnameable blues and its ability to nourish you and kill you at the same time. There was unspeakable power at Jaws, but it was the beauty that got me.

As though reading my mind, King said, "It's so rare to find water this clear next to a big wave." The clarity made for stellar imagery, and as Hamilton dropped onto the wave, King raised his camera and began to film. Now I was looking at a picture I had seen before: Hamilton, crouched in his distinctive solid stance, slicing through the barrel, brushing Jaws' molars as far back as fate would allow. He carved a few turns, slingshotting himself forward, and then as the wave snapped shut he raced back up the shoulder and launched, flipping a backside aerial, framed against the clouds. Emory zoomed in for the pickup; with barely a pause Hamilton grabbed the trailing rope and they raced back to the lineup. A minute later Hamilton appeared on another wave.

They did laps like that for hours, trading off every six waves or so. King and I skirted the perimeter, while he shot from various angles. For all I'd heard about the bad behavior and clogged conditions that had plagued Jaws in recent years, things seemed controlled. This swell wasn't remarkable enough to win anybody a prize. It was simply an unexpected gift from the weather gods.

I was sitting sidesaddle on the Jet Ski, content to watch wave after wave until the sun went down, when Hamilton drove up. "Jump on," he said, indicating his Ski.

I wasn't sure what he had in mind but I climbed on behind him, and we motored back to the lineup, where Emory waited in the water with his board. Hamilton turned and examined my flotation vest, checking that it was cinched tight and fully operational. Then he threw Emory the tow rope, nodded at a set on the horizon, and hit the throttle. "Let's get a wave," he said, and made straight for the takeoff zone. I tensed my legs, hugging the Ski, and I tightened my grip on a seat strap that suddenly seemed very flimsy. Hamilton stood, looking over his right shoulder at Emory and at Jaws building behind us, the Jet Ski pinned at full speed. Emory cut back through the wake and snapped away from the rope, dropping it. Now we were on the wave itself, near the top of the face as it began to stand up, and I knew that any second Hamilton would exit stage left and peel off the back, circling around the side for the pickup.

Except he didn't.

Instead, he held the line. I realized with a shock that we were soaring straight down the face of the wave: in effect, surfing Jaws. Emory was so close I could see him looking at us, his eyes wide with surprise. In a spasm of violence, the wave jacked up. It pressed us forward as we dropped down a wall so steep that I felt sure I would get pitched over Hamilton's head. Emory veered right and Hamilton glanced up at the lip that now towered above us, calculating exactly how many seconds we had before Jaws swallowed. The G-forces made it hard to turn my head, but at the edges of my vision I saw spray and froth and the blood beat hard in my ears as the wave bellowed only yards behind us. We'd gone left—Hamilton and Kalama's specialty. Hamilton shot forward in a burst of power and we outraced the falling lip, rocketing into the impact zone, heading directly for the inshore rock field.

He whipped the Ski around, avoiding the obstacles and neatly swinging across to where Emory had exited. "You don't need a surfboard to surf, you know," he said, smiling. "Up for another one?"

It was a rhetorical question and Hamilton, above all people, knew that. Every cell in my body vibrated. Did I want another wave? I wanted another *ten*, and then another ten after that. Though it would be weeks before I fully processed the feeling of riding Jaws, nothing I had ever done or seen or been through had made me feel so alive. Intellectually, I had always known that big-wave surfers were addicted to their pursuit. Now I knew why.

"The swell energy was good on this one," Hamilton said, opening up a dashboard compartment in the Ski and extracting a granola bar. "Long interval." King and I sat in the channel with him as the afternoon wound down. The waves still had plenty of bounce, but the surfers were done for the day. Their eyes were bloodshot, their throats were raw from shouting, and they could hear that inner voice reminding them, *The worst injuries happen when your guard is down.* Emory had gotten a lift back to Maliko, and King was interviewing Hamilton on camera. This was the best possible time for questions. After a thirty-wave run, Hamilton was relaxed and at his most voluble.

Every swell came in with a different tempo, Hamilton continued, explaining how sometimes the energy was clean and organized while at other times the waves were dangerously erratic. They could build from different angles that acted at cross-purposes: "When that happens it's really easy to get yourself into a situation where if you fell, another wave would be right on top of you." Reading the ocean's nuances was a key skill for big-wave riders, and it took years to develop the highest levels of sensitivity. Waves that looked alluring from the cliff could turn out to be fraught with quirks and pitfalls and the kinds of surprises you really don't want to encounter on a fifty-foot face. By comparison, today's waves had been quite well mannered.

If you were truly in touch with the elements, the riders claimed, you could not only see the waves' rhythms but *feel* them. "Your senses can get attuned to the most minute things going on in the water," Dave Kalama had said. During the Halloween swell in Tahiti I had tried an experiment along these lines, slipping into the water and swimming toward Teahupoo's shoulder. I wondered if it was possible to pick up its energy as it blasted through the sea. My venture was abruptly halted by the captain, Eric Labaste, who'd gestured angrily for me to get the hell back in the boat. *"Non, non, non!"* he said, shaking his head as I climbed on deck. *"NON."*

"I wanted to see if I could feel the energy of the wave," I had explained sheepishly to Sonny Miller, standing nearby.

"Well," he said, with a snort. "I think what would probably happen is that you wouldn't feel it at all—and then you'd feel it a whole lot."

Hamilton, Kalama, and Lickle had studied the Polynesian tradition of wayfinding, the art of using one's senses to navigate long ocean journeys. "The Hawaiians were in tune with everything," Lickle said. "They saw it. They felt it. They charted it."

Hamilton agreed. "They could watch the swell come in and say, 'Okay, the storm that generated this lasted three days.' They could see the different rhythms, the multiple beats, all the weird little characteristics. They looked at a wave and saw a complete story. They could see organization even within the chaos."

It was strange but true: in the same way an individual had his moods and habits, so did storms and waves. Thinking back on the most memorable waves I'd seen, I realized that the surfers were right—each one did have a distinct character. Jaws *was* the Grand Empress, it was that entrancing and fierce. True to reputation, Teahupoo was a wood chipper. Mavericks was a trapdoor to the dark side, and Todos Santos was a rollicking weekend in Baja, fun that could turn bad in an instant. Ghost Tree was a jagged chunk of glass, glittery in the sun, but if you handled it the wrong way it would cut you badly. Cortes Bank was a moon landing, exotic and alien, and Egypt, it seemed, was a sphinx on the prowl. They were an all-star cast in Nature's great drama, but for every wave that anyone recognized, there were infinite unknowns. We might lay eyes on them only rarely, but I knew now that giant waves were the opposite of singular events. They sneaked up on ships and hurtled by in the night and sprang to life on the far side of nowhere, seen only by satellites. In the vast, unsurveyed oceans they were always out there, racing toward an uncharted finish line, as uncountable as the stars in the sky, as present as your next breath.

"So this is Egypt?"

Hamilton and I floated, two miles offshore. The sea was a van Gogh painting from the later agitated years, livid brush strokes of blue and green and white. The sky had filled with clouds, their colors deepening to dark gold, rose, and gray. Against them, on the horizon, a frothy line of waves seemed almost to glow. I had asked him to drive me out here on the way back to Ilima's, and so we had passed Baldwin Beach and continued out beyond Spreckelsville, across the shallower reefs with their trickster jabs and quick furies, skirting the open ocean until Hamilton slowed to a crawl and said, "This is it." The Ski rocked in the swells. There was a thicker, more muscular vibe out here, a distinct frontier. Today's conditions weren't big enough to make Egypt break, but I could still get a feeling for the place. And it made me uneasy.

Hamilton gestured at a stretch in front of us. "This is the takeoff zone," he said. Trying to gauge distance, I looked toward shore, at the needle peaks of the Iao Valley brooding. A jet lifted off from the airport, making a graceful arc as it lofted into the sky. "So you fell just inside?" I asked. He nodded. "Probably four hundred yards. I'm not sure. I was so focused on making the drop—it was so long and so steep—that I didn't have a chance to see what the wave was doing. I couldn't even look down the line to make a turn. I was just trying to keep everything together to go straight."

I stared at him. Considering what Hamilton had done in his career, the waves he had ridden and the seeming invincibility that went along with that résumé, it was jarring to imagine him pushed to the edge, and to think about how far out that edge must have been. To have challenged him that much more intensely than Teahupoo at its most freaky or Jaws at its most mammoth, Egypt's waves must have been surreal. I mentioned this. Hamilton nodded, his face serious. "It was everything I could do," he said, speaking slowly. "The thing I think was the most surprising, the most different than anything I've ridden, was the speed. It was just so much faster."

Ten weeks had elapsed since December 3. Looking at Hamilton now, and hearing the tone of his voice, I realized he was still living through that day in his mind. Grisly as it was, Lickle's injury would heal; the psychological imprint would take longer to fade. As if thinking the same thing, Hamilton added, "That wave ran us down like we were going backward at fifty miles per hour." He frowned. "There was a lot of emotional stuff that went on, once we got mowed."

"What do you remember the most?"

He was silent. A tern wheeled overhead, and then dipped toward the water. The light had ebbed, turning the surface a deep slate navy. A moment passed.

"When I saw Brett's leg," he said. Pain flashed in his eyes. "It looked like a smashed orange. Just torn to the bone. And I thought, '*Ohhhhkay.*

What am I gonna do a tourniquet with? Maybe the strap off the flotation?' But then I realized, my wet suit was perfect. It was a long arm, thin, one millimeter. Which I don't always wear—but that day I did. I took it and did a double wrap, tied a knot, double wrap again, and tied another knot. And just made that thing frickin' solid. But not too tight because you can overtighten a tourniquet and fuck people up." He nodded. "Yeah, you can make them lose their leg if you leave them on too long or too tight. Then I gave him my flotation so he was lying on mine and wearing his. And then I'm like, 'Now, where's the Ski?' Then I saw the Ski, it was so far off—"

Suddenly Hamilton's voice cracked, the emotion spilling out. "I knew I couldn't swim Brett out. I told him, 'I'm gonna go . . . you have to hang in there.' So I went, and all I could think was, *'God, please don't let him bleed out.'* "

I was quiet, listening. It was one thing to hear the details of the accident and another to glimpse its personal toll. Hamilton paused, trying to collect himself. "I just remember that the one thing I was thinking about was . . . to uh, not uh . . ." He looked down at his hands. "I didn't want to have to explain to Shannon and the girls that Brett wasn't coming home," he said haltingly. "That Brett didn't make it back." Several tears rolled down his face. He reached up to wipe them away, squinting hard at the far-off shore. Then he shook his head slowly. "That was just a real day of 'DON'T get confused,' " he said, translating the ocean's rules of engagement. " 'DON'T think you're going to be able to ride everything. That's not happening.' "

"But that *was* the biggest wave you've ever ridden, right?"

The question was a gamble, given how much Hamilton hated the notion of sizing up waves that way, but I had to ask. He was silent for another long moment and then he inhaled deeply, as though steeling himself for some unavoidable chore. "Was it a monster, ridiculous thing?" he said in a quiet voice, turning to look at me. "Yes. There was one wave in particular I remember—it was like you stopped knowing how to measure.

I was counting the lips. *One—one thousand. Two—one thousand . . .* They were falling between three and four seconds. Simple math: at thirty-two feet per second, per second, it takes almost four seconds to fall one hundred twenty feet."

I looked out at Egypt, trying to summon up the image of a 120-foot wave by multiplying the day's waves at Jaws by three. I couldn't. "I don't even think a picture would give you the full—"

He cut me off. "The great thing, the whole irony of that day: there *are* no pictures." He laughed, but there was no humor in it. "Which is perfect! That one was for us. We have it. No one can take that away from us and it doesn't fade and we don't need a digital copy of it. None of that. We have it. We have it right here." He touched his chest. "It was about the experience of that day. It's like when Don said to me, 'I don't need to go out there.' And I said, 'Yes, you do. You have to. To *see* it. It would be like Tyrannosaurus rex is right over the ridge and you can go look at him and he's eating something! You *have to* go look.'"

"I still think the weirdest thing was that nobody expected those waves to be here," I said. "How do you think that happened?"

Hamilton cupped his hands so they were making a letter V. "It's like a light. When you're closer to the storm, the window is a lot smaller, and a lot more intense. I think the storm was funneled right at us. We were right in the center of the energy. And why wasn't it forecast? Well, there's no buoy that would have captured it. There is nothing positioned to the north and west of us." He shrugged. "I also think they don't know how to read storms that close. They're not used to figuring out how big the surf is going to be one hundred miles off the eye."

The wind had picked up and we had drifted even farther out, so Hamilton reached down with the lanyard and started the ignition. As he did so, two humpback whales appeared about fifty yards away, their knuckled backs skimming the surface. They spouted with a faint huffing sound. We were in the lair of giant waves and mammoth sea creatures, and it was a good idea not to forget that. "There's some power out here," Hamilton said, watching them pass through the swells. "But that's what it's about. That power."

His remark stayed with me as we turned away from Egypt. Around us the waves were breaking and tumbling and churning like the restless auguries of some distant storm, but at the same time everything felt peaceful. Like the sea, we are always in motion. The waves loom in our dreams and in our nightmares through all of time, their rhythms pulsing through us. They move across a faint horizon, the rush of love and the surge of grief, the respite of peace and then fear again, the heart that beats and then lies still, the rise and fall and rise and fall of all of it, the incoming and the outgoing, the infinite procession of life. And the ocean wraps the earth, a reminder. The mysteries come forward in waves.

EPILOGUE

On satellite photos Super Typhoon Nida was a fearsome sight, a huge, menacing vortex with a perfectly circular ink-black eye. A sign of brute force, Nida's eye was so wide and well-defined that fifty-foot seas were clearly visible through it, and as the storm swept northwest, loosing 170-mile-per-hour winds, giant waves, and torrential rain on the islands of Okinawa, Guam, and Yap, weather experts ventured that this was the most intense cyclone ever to develop in the month of November, and one of the strongest on record.

She lit up the Pacific, this Queen of the Magenta Blobs, weakening slightly to a category four, only to roar back again as a mighty five. As she dervished toward the Aleutians, wind shear interference slowed her down, and she devolved into a less powerful—but still sprawling—low-pressure system. When Nida's tropical moisture collided with other, colder storms in the area, Sean Collins noted on Surfline that "this is like throwing twenty gallons of gas on an already raging bonfire." What surfers were in for, he wrote, was some waves for the ages: "Our December 2009 storm will be a very special one."

By December 3 the radar maps were awash in purple. "The entire Pacific is in an uproar," Surfline reported. The result of all the turmoil was a beast of a northwest swell aimed squarely at Hawaii. Its waves would be massive, the forecasters reported, possibly even eclipsing those of December 4, 1969, when the largest swell in Hawaii's history hit the islands, washing away roads, dragging cars out to sea, depositing boats far inland, and ripping houses off their foundations. "As Hawaii's Seas Roil, Surfers Await the Big One," read a headline in the *New York Times*. "Climes Against Humanity as the Weather Goes Wild," trumpeted the *New York Post*, "50-Foot Waves Hit Hawaiiieee!" and "Super Swell Coming," warned the front page of the *Honolulu Advertiser*. "Oahu Braces for Monster Surf."

Dave Kalama, monitoring the swell as it bore down on Maui, was shaken by its size and intensity. "It was the biggest storm I've seen on the forecasting maps *ever*," he'd tell me later. "It kind of sent me into a tailspin. I had to think about how much I wanted to risk."

Hamilton, who was training on Kauai, heard news of the waves through various channels, none of which involved a computer. "I try to avoid looking at the screen as much as possible," he said. "I listen to what everyone's saying and then I end up having my own opinion." Always skeptical of the hype that attended a potentially serious swell, in this case even he was wary: "I was gearing up for the unknown."

After forty-two years in Hawaii, Hamilton had vivid memories of famous storms. The ocean, as he had witnessed, was capable of far grander furies than people consciously allow for. He was five years old during the epochal 1969 swell, and recalls his family's midnight evacuation from their house on Oahu's north shore. "I remember getting into my dad's 'fifty-six Chevy convertible," he said, "and waves were running under the car, under my feet." In 1992 he was living on Kauai when Hurricane Iniki tore through, destroying more than four thousand homes (some of which were carried out to sea by forty-foot waves) and leaving the island in tatters. On another childhood occasion he remembers watching Kings' Reef break, a deepwater wave outside Hanalei Bay that appears so infrequently as to seem mythical. To erupt into its full splendor, Kings' needs a rare—and monumentally powerful—set of conditions that happen only once or twice a decade. Hamilton had never forgotten the sight, or the size, of Kings' Reef. "It's always in the back of my mind," he said. "You could get a two-hundred-foot wave there. One hundred, easy."

As the swell created by Nida advanced on the islands, however, it became clear that the wave most suited to its direction was Jaws. "It looks like a perfect Pe'ahi swell," Hamilton said. "And we haven't had one for so long." He gathered up his gear and flew back to Maui.

The sky was restless the night before the waves arrived. I drove through Paia, the town twinkling with Christmas lights, and then up the Hana Highway under a strange horizontal half-moon. Thin clouds scudded in front of it, muting its amber light. There were few cars on the road, a general air of desertion. At Hookipa I pulled over and got out. A damp blast of air and a wall of ocean noise rushed at me, the rising energy from the first creeping edges of the swell. Even in the dark I could tell the waves were coming up. White crests flashed boldly offshore, and the pounding bass of the break was loud and insistent.

Across the island people had spent the day mobilizing for the approaching high surf. At north shore beaches the lifeguards moved their towers inland. County workers coaxed the homeless back from shorelines and advised residents whose property might be underwater by morning. Boats returned to marinas. Launches were closed. Rescue services added personnel. "How about these waves we've got coming?" a cashier said to me at the grocery store, his face beaming with excitement.

Hamilton and Kalama had been busy prepping Jet Skis, tow ropes, boards, and rescue equipment. "The bigger it is, the more redundant we get," Hamilton said. "We've got a checklist of things we go through every time. Did we change the plugs? Tighten our footstraps? How are the sleds? Do we need oil? Radio batteries? There are just multiple layers of stuff. I'll go through it. And then Kalama will go through it. Then I'll go back through it again. And then *he* might go through it one more time. It's a ritual of consequence."

The board Hamilton planned to ride was one he called the Green Meanie. Handmade by master shaper Dick Brewer, the Meanie was a sleek six-foot spear with a cruel point and razor-sharp fins that, according to Hamilton, "had some magic in it." Not every board did. Often he had to test a dozen to find one that really clicked. "Some I can just *look* at and know I don't like them," he said. "When the energy isn't flowing quite right."

Along with its spiritedness, the Meanie could handle speed. "It's got a fifth gear," Hamilton said, adding that the last thing a person wanted to feel when he was making a bottom turn on a seventy-foot wave was his board chattering around beneath his feet. But like any twitchy thorough-bred, the Meanie occasionally misbehaved. It had been present four years earlier when Hamilton buckled his knee in Tahiti, and at Egypt when Lickle's leg was flayed. (Lickle, in fact, believed the Meanie's tail fin had been the instrument of his injury.) Though surf lore held that green boards carried bad luck, Hamilton shrugged off the concern. "When they work like this one, you don't get rid of them too quickly."

The Meanie had been to Jaws before, but never on a swell of this size. After all, four years, fifty-one weeks, and a day had passed since the last one. In that regard this storm was a reunion, the first time Hamilton's full crew had been together since December 2004. In the late afternoon Darrick Doerner arrived from Oahu, along with Jamie Mitchell, Don King, Sonny Miller, and two camera assistants. Big-wave veteran Terry Chun flew in from Kauai; along with Doerner and Mitchell he'd trade off riding waves with running safety for Hamilton and Kalama on another Jet Ski. Don Shearer, patrolling overhead in his helicopter, and a pair of medics who'd be stationed on a boat in the channel, rounded out the team.

But there was one person missing.

Earlier in the day I'd spoken to Brett Lickle's wife, Shannon, and asked if he was planning to be out on the water. "He's not sure," she said. "But . . . I don't think so." In the two years since the accident Lickle had sat out several swells, but one time in 2008 he'd agreed to get back in the saddle and tow Hamilton. The session had ended badly, with Hamilton frustrated and Lickle feeling gun-shy and rattled. Despite this, I wondered if the sheer drama of this event would make him change his mind. Only the morning would tell.

In the predawn shadows Kahului Harbor bustled with riders, photographers, boat captains, and all the heavy machinery that accompanied them. The first thing I saw when I pulled into the parking lot was a truck and its trailer being sucked into the sea by waves that licked the top of the launch ramp. The driver's eyes bulged and his tires spun as he gunned it on the slick cement, but he was headed backward fast. The ocean pawed the top of his sidewalls. It took quick action with a winch to save him; luckily there were many around. Standing next to the pier, a tow team I didn't recognize zipped themselves into double flotation vests, expanding into aquatic Michelin Men. As I walked by them I caught the twinned scent of marijuana and fear.

Jet Skis and boats backed up to the fast-rising water, sliding into the harbor. All other access points on this side of the island were closed. "This is the real deal," Don King said, looking out at the ocean. "And the tide is super-high." He stood on the deck of our boat, a thirty-five-foot vessel called the *Kai Kane* (Hawaiian for Water Man), captained by local fisherman Alan Cadiz. Hamilton, Kalama, and the others were just ahead of us on four Jet Skis; Shearer would lift off from the heliport with Miller at first light.

Jaws was likely to be crowded; the pent-up desire to ride the wave at full throttle guaranteed that. "Everybody wants one," Kalama said earlier that morning. "Watch—at least half the guys out there today shouldn't be." The potential for chaos was heightened by the fact that many experienced riders would be missing from the roster. Men like Greg Long, Twiggy Baker, Carlos Burle, Garrett McNamara, Kealii Mamala, and Shane Dorian had remained on Oahu for the Eddie Aikau competition.

Known as the Eddie, this was paddle surfing's most iconic event.

Held to honor Aikau, a Hawaiian waterman lost at sea in 1978 while pad-dling for help after he and his teammates became stranded on an outrigger canoe, the Eddie was part big-wave contest, part hallowed memorial. Adding to its aura, it rarely took place: since its inception in 1984 it had been held only eight times, when the waves at its venue, Waimea Bay, were pristine and topped twenty-five feet. This year it looked as though the weather gods had finally delivered. The only question: Would the waves be too big?

Hamilton and Kalama—both of whom would rather eat glass than enter a surf competition, even the Eddie—were in pursuit of something entirely different. The *Kai Kane* chugged out of the harbor and into the rollicking Pacific. Cadiz, an athletic guy in a navy sweatshirt and surf shorts, stood barefoot at the wheel. He surveyed the ocean with height-ened alertness, as if the water's every movement carried some greater portent. Today was not a day for carelessness.

The leading edge of morning came, gauzy yellow light against a tab-ula rasa sky. It was hard to predict what the weather would do. With such a formidable storm parked (relatively) nearby, anything was possible. "Every swell is different," Don King said, pulling up the latest buoy readings on his phone. Cadiz nodded agreement, flicking on the radio. Static crackled and the computer voice that delivered the marine weather boomed through the cabin: *"A large and dangerous northwest swell will continue to bring very large surf to the islands through at least Wednesday,"* the voice said in its flat machine drone. *"Surf will exceed warning levels . . . some areas may experi-ence waves in unusual places."* Today was Monday and that swell was here all right, galloping past us in bus-size berms with deep troughs between them. The boat lumbered up and down in a slow, purposeful rhythm.

We passed outside Hookipa. The beach entrance was closed, cor-doned off by police. A wall of spray obscured the shore, waves exploding on the outer reefs. I looked up at Haleakala. The sun flared off the obser-vatory at the volcano's peak, making it gleam like a beacon. Jaws lay ahead. Cadiz swung wide and eased the engine, scouting the scene before venturing closer. King leaned over the rail and smiled. "It's *on*," he said. In the distance I could see the ocean spitting long plumes of white from what appeared to be a soft-edged, surging mountain.

Carefully we made our way toward the channel. Jet Skis flitted in from all directions. Three helicopters flew tight circles at low altitude; one had a movie camera mounted on its nose. The wave seemed to form from miles out, building from bump to peak to howling monster. We were nearer now, and the frenetic energy began to settle and focus. "When you get close enough to actually hear it and feel the noise in your chest," Cadiz said, "that's when you *realize*." King stood on the back deck and pulled on his wet suit and flotation vest. I did the same, though my hands were shaking and I was far from sure I wanted to go in the water. As we idled in the channel, Cadiz examined the waves' angle. Even a slight twist to the north could swing the break in this direction.

"Ohhh, look at this set!" King said, pointing. The wave rose up, its face veined with white, and it lunged against the sky and hung there, steepening and feathering, before the lip knifed down and burst the surface, unloading for a deafening three seconds. And then everything was white, boiling water, shimmering air, the liquid shards of a shattered mirror, and the *Kai Kane* rocked in the aftermath. This was a different Pe'ahi than I'd encountered before.

Don Shearer's yellow helicopter swooped over the cliff and across the water, pulling in front of the barrel. Shearer flew so low that at times the wave's crest was above him. His movements were so precise and intuitive that his helicopter could trace the path of the rider, close enough to monitor his facial expressions. Miller, strapped in a harness, leaned out of the doorless chopper, filming. When a board or a Jet Ski or a surfer was lost in the whitewater, Shearer would spot its location from above and then hover directly over it, signaling the rescuers.

Out of the corner of my eye I saw a Jet Ski approaching. Doerner, wearing a red windbreaker and surf shorts, with a pair of rescue fins strapped around his waist, pulled up alongside. He was a normal-looking guy with abnormal intensity, and though I'd never watched him in action before, I had the instant impression of deep expertise. Doerner often unleashed his sharp wit against the clueless flailers of the tow-surfing world (though he was always prepared to rescue them). "This is the Sport of Kings," he had said. "Not the Sport of Bozos." Today his dark brown eyes flashed the message: *This wave is serious, and anyone who's out here had better damn well be serious too.*

Cradling his bulky camera housing, King slung a leg over the gunwale, then turned back to give me some parting advice. "If you get hit by a wave," he said, "the thing is not to panic. You don't want to stay completely loose—that's how things get dislocated—you want to kind of ball up. Remember: you *do* have the lung capacity to wait it out." His voice sounded as though it were coming from a great distance, as the roar of the wave warped the noises around us, gulping them up. Then he jumped onto Doerner's Ski and was gone.

Five minutes later Hamilton shot up to the boat to get some water from the cooler on deck. "What's it like?" I asked, handing him a bottle. "It's all you want," he said. "You can bite off as much as you want. You'll get an idea of the energy . . ." The rest of his sentence was swept away by the thunder of another breaking wave. Then he circled around and yelled something I didn't catch except for the words "out there." He left as quickly as he had come, but Terry Chun arrived next, and he motioned me onto his Ski.

The massive waves stayed for two days. Even after that, for the rest of the week, the seas remained unruly. If it wasn't the historic swell that had been heralded—the demon storm that would wipe Maui off the map—it was still something very good and very rare. In celebration of the waves, as the dusk began to settle after Day One, the tradition resumed of gathering on the cliff to toast Pe'ahi.

Trucks and golf carts bumped down through the pineapple fields:

Hamilton, Kalama, Mitchell. Sierra Emory was there, and so was Teddy Casil. Miller, King, and the French photographer Sylvain Cazenave brought their cameras. I saw Lickle standing against a flimsy fence that someone had erected, and I walked over. He was watching a lone tow team through binoculars, still out on the wave; a die-hard duo that hadn't yet learned how Jet Skis break down constantly, and when that happens it's unfortunate if you're out in the storm-tossed Pacific Ocean at night. "It's still got size," he said as I approached. "And my prediction is that it'll be bigger tomorrow."

"Having any second thoughts?"

Lickle lowered his binoculars and shook his head. "The way they hyped this swell, I was so stoked *not* to be involved. I mean, I'm already broken, so when they start going 'THE BIGGEST WAVES IN A HUN-DRED YEARS!' I'm thinking, 'I don't want to be there.'" He laughed. "I just want to be here, on the cliff."

From this vantage point Lickle had watched the scene unfold in full Technicolor, and it had given him much occasion to feel grateful for the solid ground beneath his feet. Three Jet Skis had been lost to the rocks, along with multiple boards and a yard sale's worth of gear. Though Hamilton, Kalama, Mitchell, Emory, Shearer, and others had expended no small effort trying to help people who had gotten themselves into tight situations, there were several close calls. Australian rider Jason Polakow barely survived a three-wave hold-down after being sucked over the falls and having his flotation vest blown off. He'd been pinned below—among the crevices—for a full minute, an eternity in that kind of turbulence, and when he finally emerged on the surface, his lungs were half-filled with water. "I am so lucky to be here," he said afterward, his face ashen. "I saw myself dying. I could feel my brain just systematically shutting down."

Out on the Ski with Chun, I had seen a rider get struck square on the head by the lip of a wave, then disappear for a painful duration. "That's like being hit by a car," Chun said. "He's in darkness right now. I hope he's all right. But I doubt it."

There was a steady stream of carnage. Inexperienced drivers had been caught inside, Jet Skis trapped below the peak; tow lines had crossed, causing high-speed collisions; shaky riders had dropped into waves that had already begun to close out, doomed before they started, erased in a cannonade of whitewater. Midafternoon a front had moved in, the temperature dropped, the wind changed, and with it the rules of the game. "Onshore wind," Hamilton pointed out, watching the latest casualty, a guy in a neon green wet suit, his face covered in white zinc oxide, bouncing backward down the face of Jaws. "He's gonna be under there for a while."

The circus, the jeopardy, the nerves—Lickle made it clear he didn't miss them. "I actually slept last night," he said. "I wasn't like some spun monkey, getting up every hour in a cold sweat." I believed him. Yet I wondered if a person with a taste for the edge would ever be fully content on the sidelines. There was always something bittersweet about passages, the awareness that time turns everything, even the most reliable truths, into talismanic memories. While Hamilton and Kalama had dominated the lineup on this day, just as they always did, a post-big-wave era had arrived for Lickle—and mixed in with his obvious relief, I sensed a dash of sadness. I asked him if this was true.

He answered quickly, as though he'd already given the question considerable thought: "The only thing I'll say is that the accident was kind of a ticket out, you know what I mean?" His voice was gruff but full of emotion. "What we had was a *gang*. And you couldn't get out of the gang. There was no way out. There's so much peer pressure like, 'Come on, you're the man! Let's go!' You can't just walk away because . . . you just can't. But if you get shot up and almost die, they let you out."

It was a heavy price to pay, but that was why it mattered. Though Lickle's time in the big-wave arena was over now, that didn't take away what had been. "Looking back on that day," I said, "knowing what you had to go through, would you do it again? Was it worth it just to surf that one wave?"

"Absolutely," Lickle said emphatically. "I was the highest I've ever been when I got off that rope and got on the Ski to tow Laird." His face turned grave. "Then I went from there to a place that was so low, basically bleeding to death. But oh, yeah, even if I knew . . . I still would've ridden that wave."

I glanced down at his left leg, the scar so prominent that it seemed as though his entire calf had been melted. The flesh was scrambled, annealed. Lickle raised his binoculars again. "There might be a little more north in the swell right now," he said, brusquely changing the subject.

Behind us things were raucous, the red wine and Coors Light flowing. "You're cut!" Kalama shouted affectionately at Don Shearer. "You're making us look bad! We used to be the ones looking cool in the waves! Now it's 'Hey, who's the guy in the helicopter?'"

Miller, standing nearby, let out his trademark infectious laugh. "Our band hasn't played in a long time," he said, and gestured toward Jaws. "It's like the sleeping giant has awakened!" He leaned against the bumper of Hamilton's truck. "There are some crazy waves that people are getting into, Tasmania and wherever, but on a day like today you're reminded— Pe'ahi is the master of them all."

Hamilton, overhearing, agreed: "When she's on, everybody better have their heads bowed."

In recent years tow surfing had opened a new frontier, the "crazy" waves to which Miller referred. Called slabs (or sometimes death slabs), they were more like oceanic car wrecks than proper waves, as thick as they were tall, fractured and brutally misshapen, with cavernous, sucking holes at their bases. They formed around reefs and ledges where strong ocean swells were forced abruptly from deep to very shallow water, leaving a rider no margin for error. "Some waves are walls and some are ceilings," Hamilton had said once, talking about Teahupoo, which was at heart a slab. But that wave had been to finishing school compared to some of the slabs off Australia's coast, which seemed to have come from Nature's insane asylum. "It's ridiculous," Mitchell told me. "It's a whole different deal. There's a line. Fall on one side of the line, you're fine. Fall on the other, you're dead."

The raggedy slabs held little interest for Hamilton, though other wave frontiers did. While age had blunted some of his sharper edges—his regular 120-foot cliff jumps, for instance, were now a thing of the past—in no way was he slowing down. "I'm in it for the long haul," he'd said, adding that experience was exactly what he needed to progress. "You're a better gambler when you can afford to pay the bill."

Still, there were things he disliked—the flash and hype that now surrounded tow surfing at the top of that list—and I suspected that in the future he would try to avoid them. One means for doing so was the hydrofoil surfboard, a hybrid device he'd jury-rigged about a decade ago and had toyed with ever since. Foiling, as it was known, was an odd-looking pursuit where the surfer floated four feet above the wave, clad in snowboard boots. The foil board itself was even smaller than a tow board, steered with an underwater rudder connected by a vertical strut. As ungainly as it appeared, hydrofoil surfing allowed riders to carve graceful arcs through even the choppiest waves. All friction was gone.

But some kinks remained in development. Falling, for instance, was a fast route to serious trouble. It was one thing to eject from a pair of foot straps, and another to unbuckle snowboard boots underwater while getting whipsawed by the contraption itself. The steel hydrofoil—the wing that sliced through the wave, just below the surface—was heavy and pointed, capable of terrible damage. In one mishap, Don King, shooting underwater, had narrowly escaped decapitation.

Regardless, Hamilton was optimistic. "I think foiling will evolve to help us get through the barriers on giant surf," he said. "It will allow us to go to the next dimension, which is faster. We won't be affected by surface conditions." Force Twelve storms in the North Sea, dangerously raw swell at the edge of a hurricane, waves too enormous or chaotic for towing—with the right equipment, all of these rides would become possible: "Ultimately the objective is to ride the biggest swells the ocean can create."

These new extremes that Hamilton sought? He was likely to find them. Everything in the oceans, it seemed, was rising: wave heights, sea levels, surface temperatures, wind speeds, storm intensities, coastal surges, tsunami risks. "Now is the Time to Prepare for Great Floods," a July 2009 editorial in *New Scientist* magazine advised, predicting that "great swathes of urban sprawl will vanish beneath the waves" as the oceans creep higher. "It's easy to imagine an apocalyptically soggy future for New York," warned *New York* magazine, "high waves soaking the hem of Lady Liberty's robes, flash floods roaring through subway tunnels, kayakers paddling down Wall Street." "The future of the UK's coastal cities is in jeopardy due to rising sea levels," Lloyd's of London reported in one of its bulletins. "The Bigger Kahuna," read a recent headline from *Scientific American*. "Are More Frequent and Higher Extreme Ocean Waves a By-Product of Global Warming?"

The relationship among the waves, the weather, the planet's rising temperatures, and the overarching ocean cycles is wildly complex—and our understanding of it is far from complete—but the short answer is: almost certainly yes. "The increases are important in their impacts ranging from ship safety to enhanced coastal hazards, and in the engineering design of ocean and coastal structures," researchers at Oregon State University concluded. In a hallmark paper, they had just revealed that the hundred-year wave height in the Pacific Northwest, measured at 33 feet in 1996, was now closer to 46 feet, and by some calculations might even top 55 feet.

It wasn't hard to imagine seas that size—and larger. All I had to do was look over the fence. At Jaws the waves just kept coming. Whitewater billowed and tumbled from the break to the cliff, the energy that had surged across the Pacific coming to an end on these rocks. Hamilton walked over to Lickle, trailed by Buster, his rat terrier. "It's still bombing," he said. "And the buoys are up."

"Well, if you look at the blob," Lickle said, "it's still here—and it's still purple. It's not leaving. It's just sitting here." He nodded sagely.

"There were some big waves today, but they were probably just the front-runners. I think tomorrow's when we're really gonna get it."

Hamilton leaned across the fence, suddenly focused on the water. "Some of these waves are unbelievable."

"Hopefully it doesn't peak tonight," Lickle said. "Hey, just for the record," he added in a low voice, as though he were about to reveal a secret.

"Yeah?" Hamilton looked up.

"That one wave almost took you out." Lickle chuckled darkly.

"Which one?" Hamilton thought for a moment. "Oh . . . yeah. *That* one. The paint stripper."

"You were ten feet under the lip!" Lickle's tone was incredulous. "The fact that you made it through at all . . ."

"I'm glad you saw that," Hamilton said, laughing. "Submarine ride."

The air felt moist; the sky was full of moody grays and purples. A front was moving through, shifting the winds back offshore. But the gusts remained light and would likely stay that way come morning. "This day's had three days in it," Hamilton said. "That's what makes it so beauti—"

"How did you ride today?"

A tiny voice interrupted Hamilton from behind. Sky Lickle—all four feet of her—stood there, hands on hips, wanting to know.

"How did I ride?" Hamilton said, searching for an adverb. "Um— successfully?"

"You made it through!" Sky's face lit up in a smile.

"Exactly!" Hamilton laughed, his entire body radiating happiness. He leaned over to give Sky a high five, and then he stretched out his arms as if to embrace it all, the waves and the fields and the people around him. "*That's* what I'm talking about!" he said. "You know what? That sums it up: 'You made it through.'"

ACKNOWLEDGMENTS

I never expected it to be easy, examining the secrets of giant waves. It would be a dicey kind of hunt, I thought, and for any hope of success I'd need a guide. For this I approached Laird Hamilton. My gratitude to him is immeasurable: Not only was he willing to open up his world to me, providing an extraordinary glimpse of the ocean with its gloves off, he and his wife, Gabby Reece, opened up their hearts as well. Through them I learned the true meaning of aloha—the lovely Hawaiian tradition of giving deeply of yourself, even to someone you don't know very well.

I found that same generosity of spirit throughout my reporting, no matter where in the world I ventured, but especially in Maui. My profound thanks also go to the Lickle family—Brett, Shannon, McKenna, and Skylar—who provided me with my favorite home on the island, along with many wonderful conversations and dinners; Dave and Shaina Kalama; Teddy Casil and Devri Schultz; Don and Donna Shearer; Sonny Miller; Jeff Hornbaker; and Don King. On the mainland, I owe much to Don and Rebecca Wildman, and Ron and Kelly Meyer. And how can I possibly thank Jane Kachmer, an extraordinary woman without whom this project never would have happened. Her endless support, hard work, and infectious optimism are truly appreciated.

So many people helped me in the waves. In the surfing world I send a most heartfelt *shaka* to: Darrick Doerner, Sierra Emory, Gerry Lopez, Greg Long, Twiggy Baker, Sean Collins, Garrett McNamara, Kealii Mamala, Jeff Clark, Tony Harrington, Mike Prickett, Jamie Mitchell, Art Gimbel, Terry Chun, Nelson Kubach, Martha Malone, James "Billy" Watson, Mike Parsons, Brad Gerlach, Peter Mel, Ken "Skindog" Collins, Raimana Van Bastolaer, Teiva and Nina Joyeux, Tim McKenna, Randy Laine, Maya Gabeira, Ricky Grigg, Greg Noll, Bill Ballard, Josh Kendrick, Scott Taylor, Butch Bannon, Rob Brown, Tom Servais, Erik Aeder, and Sylvain Cazenave.

In the science realm my list is equally lengthy. Huge thanks to: Penny Holliday, Don Resio, Val Swail, Al Osborne, Peter Janssen, Margaret Yelland, Sheldon Bacon, Peter Challenor, Christine Gommenginger, Russell Wynn, David Levinson, John Marra, Steven N. Ward, George Plafker, Lawrance Bailey, Ken Melville, Enric Sala, Jeremy Jackson, Paolo Cipollini, Meric Srokosz, Peter Taylor, Andy Louch, Joanne Donahoe, Kim Marshall-Brown, and Mike Douglas.

In South Africa, I am indebted to Nicholas Sloane, Jean Pierre Arabonis, Dai Davies, and Desiree Bik. In London, I thank Neil Roberts at Lloyd's of London, and Bill McGuire at Aon Benfield UCL Hazard Research. Anyone interested in venturing deeper into Lloyd's of London's fascinating world will find its Web site riveting; you can get lost in there for hours. Likewise, I highly recommend Bill McGuire's books for further reading about nature's harrowing extremes.

Hawaii proved to be far more than the main location for *The Wave*. It became my home, and a place of solace after my father's sudden death in July 2008. I had barely started writing at that point, and in the aftermath I didn't even know how I could, but the people and the land and the ocean helped me through that dark time. Everywhere in Maui I encountered outlandishly giving people, and I will never forget their kindness: Rich and Ann Marie Landry (and all the Maui pink cap swimmers!), Gary Ryan, Marie Cruz, Ed and Kerri Stewart, Felice and Paul Miller, Chelsea Hill, Cheyenne Ehrlich, Ian Horswill, the Reverends Shelley and Kedar

St. John, Skeeter Tichnor and everyone at The Studio Maui, Chinta Mackinnon, Tim Sherer, Doug Fujiwara, Skip Armstrong, and Eddie Cabatu. I send a special super-aloha to Phyllis Tavares, the founder of Ninth Life, a no-kill cat shelter in Haiku. Phyl adopted the family of stray cats that took up residence on my porch, nestling together behind my surfboards. If you want to read the inspiring story of a local hero, check out her Web site: www.9thlifehawaii.org.

Back on the mainland, as well, I am grateful to many friends. Andy Astrachan read the earliest version of this work, and his encouragement buoyed and inspired me. My deep thanks also go to Hilary Laidlaw, Niccolo Ravano, David Lynch, Sharon Ludtke, Kristin Gary, Samantha Carey, Eldar Beiseitov, Ann Jackson, Susan Scandrett, Tim Carvell, Tom Keeton, Celia and Henry McGee, Susan Orlean and John Gillespie, Jill Meilus, Vic Calandra, Peggy Dold, Isolde Motley, Maria Moyer, Ace Mackay-Smith, Mark Taylor, Paula Blanchet, Susan King, John and Jane Clarke, Mary Lou Furlong, La Mura Boelling, Dr. Guldal Caba, Dr. Lionel Bissoon, and Leslie Fischer.

As always, I owe much to my family: my mother Angela Casey, my brothers Bob Casey and Bill Casey, along with Pam Manning, Beth Oman, Mike Casey, Caroline Casey, Kellie Casey, John Casey, Wilda Alford, Lorna Walkling, Tom Walkling, Chris Walkling, and Sarah Walkling-Innes.

Among my colleagues I am especially indebted to Martha Corcoran, whose help throughout this project was invaluable. Tom Colligan and Cathay Che contributed key reporting and fact checking. Sara Corbett and Lucy Kaylin read a galley version of the manuscript and offered much insight and support, as did Terry McDonell, David Granger, and Tim Carvell. At ICM I would like to thank Kristyn Keene, Niki Castle, and Molly Rosenbaum for their constant help and goodwill; and John DeLaney for his legal expertise. I owe a giant wave of thanks to my editor, Bill Thomas, whose expert guidance is present on every page, and to his amazing associate Melissa Ann Danaczko.

During the five years I spent working on *The Wave*, one person was always there: my agent, Sloan Harris. His passion and vision for this project were ever-present, and there are no words to express my gratitude for that. On any path there are ups and downs and twists and the odd obstacle and, sometimes, a lip crashing down on your head. With great humor, Sloan helped me navigate all of them.

SELECTED BIBLIOGRAPHY

BOOKS

Atwater, Brian F., Satoko Musumi-Rokkaku, Kenji Satake, Kazue Ueda, and David K.Yamaguchi. *The Orphan Tsunami of 1700*. Seattle: University of Washington Press, 2005.

Bascom, Willard. *Waves and Beaches: The Dynamics of the Ocean Surface*. New York: Anchor Books, 1980.

Bohn, Dave. *Glacier Bay: The Land and the Silence*. New York: Ballantine Books, 1967.

Bruce, Peter. *Adlard Coles' Heavy Weather Sailing*. Camden, ME: International Marine, 1999.

Bryant, Edward. *Tsunami: The Underrated Hazard*. Cambridge, UK: Cambridge University Press, 2001.

Butt, Tony, and Paul Russell. *Surf Science*. Honolulu: University of Hawaii Press, 2004.

Caldwell, Francis E. *The Land of the Ocean Mists*. Edmonds: Alaska Northwest Publishing, 1986.

Carson, Rachel. *The Sea Around Us*. New York: Oxford University Press, 1989.

Cramer, Deborah. *Smithsonian Ocean*. New York: Smithsonian Books, 2008.

Dudley, Walter C., and Min Lee. *Tsunami!* Honolulu: University of Hawaii Press, 1998.

Florin, Diacu. *Mega Disasters*. Princeton, NJ: Princeton University Press, 2010.

Fradkin, Philip L. *Wildest Alaska*. Berkeley: University of California Press, 2001.

Grigg, Ricky. *Big Surf, Deep Dives, and the Islands*. Honolulu: Editions Limited, 1998.

Hooke, Norman. *Modern Shipping Disasters*. London: Lloyd's of London Press, 1989.

Lapham, Lewis. *Lapham's Quarterly: Book of Nature*. New York: American Agora Foundation, 2008.

Lutjeharms, R. E. Johann. *The Agulhas Current*. Berlin: Springer, 2006.

Maslin, Mark. *Global Warming*. New York: Oxford University Press, 2004.

McGuire, Bill. *Apocalypse: A Natural History of Global Disasters*. London: Cassell & Co, 1999.

————. *Surviving Armageddon*. New York: Oxford University Press, 2005.

Ochoa, George, Jennifer Hoffman, and Tina Tin. *Climate*. London: Rodale Books International, 2005.

Orrell, David. *The Future of Everything*. New York: Thunder's Mouth Press, 2007.

Pilkey, Orrin H., and Linda Pilkey-Jarvis. *Useless Arithmetic*. New York: Columbia University Press, 2006.

Polkinghorne, John. *Quantum Theory*. New York: Oxford University Press, 2002.

Prager, Ellen J. *Furious Earth: The Science and Nature of Earthquakes, Volcanoes, and Tsunamis*. New York: McGraw-Hill, 2000.

Redfern, Martin. *The Earth*. New York: Oxford University Press, 2003.

Smith, Craig B. *Extreme Waves*. Washington, D.C.: Joseph Henry Press, 2006.

Smith, Leonard. *Chaos*. New York: Oxford University Press, 2007.

Smith, P. J. *The Lost Ship SS Waratah*. Gloucestershire, UK: History Press, 2009.

Sverdrup, Keith A., Alyn C. Duxbury, and Alison B. Duxbury. *An Introduction to the World's Oceans*. New York: McGraw-Hill, 2003.

Turner, Malcolm. *Shipwrecks & Salvage in South Africa*. Capetown, South Africa: Struik, 1988.

Ulanski, Stan. *The Gulf Stream*. Chapel Hill: University of North Carolina Press, 2008.

Uys, Ian. *Survivors of Africa's Oceans*. South Africa: Fortress Publishers, 1993.

Warshaw, Matt. *The Encyclopedia of Surfing*. Orlando: Harcourt, 2005.

————. *Mavericks*. San Francisco: Chronicle Books, 2000.

Williams, Waimea. *Aloha, Kauai*. Waipahu, HI: Island Heritage Publishing, 2004.

Winchester, Simon. *Krakatoa*. New York: HarperCollins, 2003.

SCIENTIFIC PUBLICATIONS, PERIODICALS, AND NEWSPAPERS

Abruzzese, Leo. "Nature's Fury." *The Economist*, November 16, 2006.

Ananthaswamy, Anil. "Going, Going . . ." *New Scientist*, July 4, 2009.

Ball, Philip. "Ship Endures Record-Breaking Waves." *Nature*, March 17, 2006.

BBC News. "Giant Wave Could Threaten U.S." October 4, 2000.

————. "Rogue Wave Theory for Ship Disaster." November 26, 2001.

Becker, Markus. "Vessel Measures Record Ocean Swells." *Der Spiegel*, March 31, 2006.

Bojanowski, Axel. "Study Sees North Sea Tsunami Risk." *Der Spiegel*, October 10, 2006.

Britt, Robert Roy. "Ship-Devouring Waves, Once Legendary, Common Sight on Satellite." *USA Today*, July 23, 2004.

Broad, William J. "New Tools Yield Clues to Disasters at Sea." *New York Times*, March 16, 1999.

————. "Rogue Giants at Sea." *New York Times*, July 11, 2006.

Calamai, Peter. "The Cold Truth About Rogue Waves." *Toronto Star*, December 19, 2006.

Carson, Rachel. "The Sea: Wind, Sun, and Moon." *The New Yorker*, July 16, 1951.

Chang, Kenneth. "Strongest Hurricanes May Be Getting Stronger." *New York Times*, September 4, 2008.

Dean, Cornelia. "Study Warns of Threats to Coasts from Rising Sea Levels." *New York Times*, January 17, 2009.

Draper, Laurence. "Severe Wave Conditions at Sea." *Journal of the Institute of Navigation*, vol. 24, no. 3, July 1971.

Emanuel, Kerry. "Increasing Destructiveness of Tropical Cyclones Over the Past 30 Years." *Nature*, August 4, 2005.

European Space Agency (ESA). "Ship-Sinking Monster Waves Revealed by ESA Satellites," July 21, 2004.

Fearing, Katie M., and Robert A Dalrymple. "Wave Refraction at Jaws, Maui." Paper from the Center for Applied Coastal Research, University of Delaware, 1998.

Fyfe, John C., and Oleg A. Saenko. "Anthropogenic Speed-Up of Oceanic Planetary Waves." *Geophysical Research Letters*, vol. 34 (2007).

Gain, Bruce. "Predicting Rogue Waves." *MIT Technology Review*, March 2007.

Haver, Sverre. "Freak Wave Event at Draupner Jacket." Paper from Ifremer Rogue Wave Symposium, 2004.

———. "Freak Waves: A Suggested Definition and Possible Consequences for Marine Structures." Paper from Ifremer Rogue Wave Symposium, 2004.

Heberger, Matthew, Heather Cooley, Pablo Herrera, Peter H. Gleick, and Eli Moore. "The Impacts of Sea-Level Rise on the California Coast." Paper from the California Climate Change Center, 2009.

Heller, Eric. "Freak Waves: Just Bad Luck, or Avoidable?" *Europhysics News*, September/October 2005.

Holliday, Naomi P., and Colin R. Griffiths. Southampton Oceanography Center cruise report. Southampton, UK, 2000.

Holliday, Naomi P., Margaret J. Yelland, Robin Pascal, Val R. Swail, Peter K. Taylor, Colin R. Griffiths, and Elizabeth Kent. "Were Extreme Waves in the Rockall Trough the Largest Ever Recorded?" *Geophysical Research Letters*, vol. 33 (2006).

Intergovernmental Panel on Climate Change, Working Group 1. "The Physical Science Basis: Summary for Policymakers." World Meteorological Organization, Geneva, Switzerland, 2007.

International Maritime Organization (IMO). "IMO and the Safety of Bulk Carriers." London, UK, 1999.

International Union of Marine Insurance. "IUMI 2006 Shipping Statistics: Analysis." Zurich, Switzerland, 2006.

———. "IUMI 2008 Shipping Statistics: Analysis." Zurich, Switzerland, 2008.

———. "IUMI Facts and Figures Committee Report." Vancouver, BC, 2008.

Kolbert, Elizabeth. "Outlook: Extreme." *National Geographic*, April 2009.

Komar, Paul D., and Jonathan Charles Allen. "Increasing Wave Heights Along the U.S. Atlantic Coast Due to the Intensification of Hurricanes." *Journal of Coastal Research*, 2008.

Kushnir, Y., V. J. Cardone, J. G. Greenwood, and M. A. Cane. "The Recent Increase in North Atlantic Wave Heights," *Journal of Climate* (1997).

Ledford, Heidi. "California Caught Off Guard by Tsunami." *Nature*, November 17, 2006.

Lloyd's of London. "360 Risk Project: Catastrophe Trends 1." Report from Lloyd's of London, England, UK, 2006.

Lohr, Steve. "Puzzled Scientists Find Waves Off Britain Are Growing Larger." *New York Times*, April 19, 1988.

Lovett, Richard. "The Wave from Nowhere." *New Scientist*, February 24, 2007.

Magnusson, Karin Anne, Magnar Reistad, Øyvind Breivik, Rasmus Myklebust, and Ellis Ash. "Forecasting a 100-Year Wave Event." Presentation from the 9th International Workshop on Wave Hindcasting and Forecasting, 2006.

Mallory, J. K. "Abnormal Waves on the South East Coast of South Africa." Institute of Oceanography, University of Cape Town, 1997.

Mangold, Tom. "Scandal of the Rotting Tankers." *Reader's Digest*, November 1993.

Masson, D. G., C. B. Harbitz, R. B. Wynn, G. Pedersen, and F. Lovholt. "Submarine Landslides: Processes, Triggers, and Hazard Prediction." *Philosophical Transactions of the Royal Society*, 2006.

MaxWave. "Minutes from the MaxWave SAP Meeting at BP Staines, 5th of November, 2001."

McCredie, Scott. "When Nightmare Waves Appear Out of Nowhere to Smash the Land." *Smithsonian*, March 1994.

McGuire, Bill. "Climate Change: Tearing the Earth Apart?" *New Scientist*, May 27, 2006.

———. "Global Risk from Extreme Geophysical Events: Threat Identification and Assessment." *Philosophical Transactions of the Royal Society*, vol. 364 (2006): 1889–1909.

———. "Ground Deformation Monitoring of a Potential Landslide at La Palma, Canary Islands." *Journal of Volcanology and Geothermal Research*, vol. 94 (1999): 251–65.

———. "There's a Storm Brewing." *The Guardian*, April 28, 2008.

Mercer, Phil. "Extreme Waves Worry Australia." BBC News, December 1, 2008.

Mertie, J. B. "Notes on the Geography and Geology of Lituya Bay, Alaska." U.S. Geological Survey, 1931.

Miller, Don J. "Giant Waves in Lituya Bay Alaska." U.S. Geological Survey Professional Paper 354-C, 1960.

Moss, J. L., W. J. McGuire, and D. Page. "Ground Deformation Monitoring of a Potential Landslide at La Palma, Canary Islands." *Journal of Volcanology and Geothermal Research*, vol. 94 (1999): 251–265.

Paulson, Tom. "Secrets of Tsunamis Not Easily Revealed." *Seattle Post Intelligencer*, January 17, 2005.

Peeples, Lynn. "The Bigger Kahuna: Are More Frequent and Higher Extreme Ocean Waves a By-Product of Global Warming?" *Scientific American*, February 2010.

Peeples, Lynn. "The Real Sea Monsters: On the Hunt for Rogue Waves." *Scientific American*, September 2009.

Perkins, Sid. "Dashing Rogues." *Science News*, November 18, 2006.

Rosenthal, W., and S. Lehner. "Rogue Waves: Results of the MaxWave Project." *Journal of Offshore Mechanics and Arctic Engineering*, vol. 130 (2008).

Ruggiero, Peter, Paul D Komar, and Jonathan C. Allan. "Increasing Wave Heights and Extreme Value Projections: The Wave Climate of the U.S. Pacific Northwest." *Coastal Engineering*, vol. 57 (2009): 539–552.

Slunyaev, A., E. Pelinovsky, and Guedes C. Soares. "Modeling Freak Waves from the North Sea." *Applied Ocean Research* 27 (2005): 12–22.

Taylor, Paul, Dan Walker, and Roy Rainey. "On the New Year Wave at Draupner in the Central North Sea in 1995." Presentation from the 20th International Workshop on Water Waves and Floating Bodies, Spitsbergen, Norway, 2005.

Tisch, Timothy D. "Tsunamis: A Rare but Real Marine Hazard." *Professional Mariner*, August/September 2005.

United States National Oceanic and Atmospheric Administration. "United States Tsunamis, 1690–1988." Publication 41–2. Boulder, CO: U.S. Department of Commerce, 1989.

Wang, David W., Douglas A. Mitchell, William J. Teague, Ewa Jarosz, and Mark S. Hulbert. "Extreme Waves Under Hurricane Ivan." *Science*, August 5, 2005.

Ward, Steven N., and Simon Day. "Cumbre Vieja Volcano—Potential Collapse and Tsunami at La Palma, Canary Islands." American Geophysical Union, 2001.

Witze, Sandra. "Bad Weather Ahead." *Nature*, vol. 441, June 2006.

Wynn, R. B., and D. G. Masson. "Canary Islands Landslides and Tsunami Generation: Can We Use Turbidite Deposits to Interpret Landslide Processes?" Southampton Oceanography Center. Southampton, UK, 2003.

Yanchunas, Don. "Crew of British Research Vessel Gathers Extraordinary and Dangerous Data on Waves." *Professional Mariner*, August/September 2006.

Yeom, Dong-Il and Eggleton, Benjamin J. "Rogue Waves Surface in Light." *Nature*, December 13, 2007.

WEB SITES

Weather and Forecasts

Surfline: www.surfline.com

Stormsurf: www.stormsurf.com

Wavewatch III Climate Model: http://polar.ncep.noaa.gov/waves/wavewatch

Institutions and Organizations

Aon Benfield Hazard Research: www.abuhrc.org

Billabong XXL Competition: www.billabongxxl.com

Lloyds of London: www.lloyds.com

National Oceanography Center (Southampton): www.noc.soton.ac.uk

National Oceanographic and Atmospheric Administration: www.noaa.gov

United States Geological Survey: www.usgs.gov

Characters

Arabonis, Jean Pierre: www.osis.co.za

Clark, Jeff: www.jeffclarksurfboards.com

Doerner, Darrick: www.dd-sea.com

Gabeira, Maya: www.mayagabeira.com

Gerlach, Brad: www.bradgerlach.com

Hamilton, Laird: www.lairdhamilton.com

Harrington, Tony: www.harroart.com

Hornbaker, Jeff: www.thirdeyeworld.com

Kalama, Dave: www.davidkalama.com

Lickle, Brett: www.surfball.net

Long, Greg: www.greglong.com

McNamara, Garrett: www.garrettmcnamara.com

Mel, Peter: www.petermel.com

Miller, Sonny: www.worldwavepictures.com

Mitchell, Jamie: www.jamie-mitchell.com

Prickett, Mike: www.prickettfilms.com

Shearer, Don: www.windwardaviationmaui.com

Sloane, Captain Nicholas: www.svitzer.com

Illustration Credits

Fourth position: Tom Servais

Fifth position: Courtesy of Tony Harrington

Page 8

Top: Courtesy of Tony Harrington

Bottom: Courtesy of Tony Harrington

Page 9

Top: Robert Brown Photography.com

Bottom: Robert Brown Photography.com

Page 10

Top: U.S. Geological Survey

Bottom, left: SSPL/Getty Images

Bottom, right: Bloomberg via Getty Images

Page 11

Top: The Granger Collection, New York

Bottom: Karsten Petersen, www.global-mariner.com

Page 12

Top: D. J. Miller/U.S. Geological Survey

Center: D. J. Miller/U.S. Geological Survey

Bottom: D. J. Miller/U.S. Geological Survey

Page 13

Top: U.S. Geological Survey

Bottom: U.S. Army/U.S. Geological Survey

Page 14

Andrew Ingram/*The Cape Times*

Page 15

Top: Benjamin Thouarsd

Bottom: Don King

Page 16

Top: Tom Servais

Bottom: From the bridge of the NOAA ship *Discoverer,* taken by RADM Richard R. Behn, NOAA (ret.)

A Note About the Author

Susan Casey, author of the *New York Times* bestseller *The Devil's Teeth: A True Story of Obsession and Survival Among America's Great White Sharks*, is editor in chief of *O, The Oprah Magazine*. She is a National Magazine Award–winning journalist whose work has been featured in the *Best American Science and Nature Writing, Best American Sports Writing*, and *Best American Magazine Writing* anthologies; and has appeared in *Esquire, Sports Illustrated, Fortune, Outside*, and *National Geographic*. Casey lives in New York City and Maui.

A Note About the Type

The text of this book is set in Fournier, a digitized
version of the original font cut that was part of
the Monotype Corporation historical typeface revivals
in the 1920s.
Fournier was created by the typographer and
printing historian Stanley Morison (1889–1967) and
grew out of his admiration for the type cuts of Pierre
Simon Fournier (1712–1768).